TEXAS ECLIPSES

THE UPCOMING CELESTIAL SPECTACLE COMING TO TEXAS AND ITS ROOTS IN TEXAS HISTORY

LETICIA FERRER

WRITE SERVICES PRESS

Published by:

Write Services Press, Horn Lake, MS 38637

Writeservicespress.com

Design:

Interior: Write Services Press

Cover: Savannah Castillo

Photos unless otherwise indicated: Daniel Brookshier; image permissions obtained and on file with author

Hardcover ISBN-13: 978-1-954373-08-2

Paperback ISBN-13: 978-1-954373-16-7

Digital ISBN-13: 978-1-954373-09-9

LCCN:

To all those who chase wonder and beauty
in their lives and who teach me every day
to do so for myself.

Yes, I mean you!

Mom and Dad, siblings, and my
different tribes of friends - work,
drinking, Toastmasters, umbraphile and
just hanging out friends.

Thank you for this life I share with you full
of wonder, beauty, and love.

CONTENTS

INTRODUCTION

I finally remembered to breathe.

It took extreme effort to tear my gaze from the heavenly sight to look around. Surrounding me, a 360-degree sunset, the bright light of the sun peeking out from the edges of the shadow. I looked up again, until all too soon, the sun returned, so bright I had to turn my eyes away.

This was my first Total Solar eclipse more than 30 years ago.

The same transcendent emotion overcomes me every time I stand under the moon's shadow, 18 additional times and counting. Each time, I find it impossible to accurately express my complex emotions with mundane words. There's a sense of wonder, primordial fear, profound awareness of the expansiveness of the cosmos, the gratitude of existing, lucky to be in the shadow, touched by the universe, the wonder of being ALIVE to witness the event. And yet, I still don't feel as if I'm giving the experience full justice in these descriptions. What I experienced in that first Total Solar Eclipse (TSE) has not faded in the other 18 eclipses I've witnessed thus far. It is my plan to see each Total Solar Eclipse on this planet until I move on from this life.

In this book, I hope to inspire you to experience the upcoming solar eclipses in Texas safely.

I'm deliberately using the word 'experience' instead of 'see' because being in the path of an eclipse is a soul touching, wonderous, awesome event. Videos or pictures cannot accurately convey the sensations of a Total Solar Eclipse. As you'll read and hopefully come to know, scientists and others who come to view eclipses with a plan of action for scientific research fail to capture measurements because they become so engrossed in the experience. They fumble in the unexpected dark of the eclipse or just forget their instruments and watch awestruck.

The upcoming Annular eclipse on October 14, 2023 and the Total Solar Eclipse on April 8, 2024, within six months of each other, are special eclipses for Texans. (Don't worry if you aren't familiar with the difference yet. That's what this book will answer for you. For quick reference, terms are also defined in the glossary.) This will be the fifth Annular eclipse and the third Total Solar Eclipse crossing Texas boundaries since the state's inception in 1836.

The first Total Solar Eclipse was on July 29, 1878, more about that eclipse in the chapter Why Focus on Texas. The next one on May 28, 1900, lasted less than a minute as it crossed Texas. Totality for that event was over what is now McAllen (founded in 1911) and Harlingen (founded in 1904), both little more than farming villages at the time. Thus, totality was over a sparsely populated region of Texas. The rest of Texas experienced a partial solar eclipse, which was seen in San Antonio and cloud-covered Galveston with what was described as "... a peculiar heavy gray twilight... with floods of freakish lights edging the clouds."

There has not been a Total Solar Eclipse over Texas in the more than 100 years since. The April 8, 2024 Total Solar Eclipse will be the third eclipse to cross the state since Texas's inception in 1836. Before then, there was an eclipse in the area in 1778, when Texas was still a Spanish colony.

CHAPTER ONE

FIRSTHAND, FIRST TIME SOLAR ECLIPSE CONNECTIONS

I t was 1991 when my late husband John Echols started having trouble sleeping. He was diagnosed with high blood pressure and other signs of high stress. He worked all the time running his construction sales company, keenly felt his responsibilities to his customers and employees, and generally never stopped thinking about his business. He was the quintessential workaholic.

His doctor, a very nice old man who was still working into his 80s, asked my husband about the last time he'd taken a vacation.

John told him, honestly, he'd just gotten back from Las Vegas. A little prodding and the doctor learned the full truth. John was in Vegas for the World of Concrete convention. His activities there included working a booth, taking clients out at night, and visiting with his vendors, customers, and friends he'd made in an over 20-year career. In other words, the normal work one does at a convention. Definitely *not* the restful activities associated with a vacation.

"TAKE A VACATION!" John's doctor ordered.

That evening, a segment on the news covered the Total Solar Eclipse happening in three weeks over La Paz, Mexico. It was supposed to be the longest lasting eclipse of the century. While watching that news segment, a song popped in my head.

According to rumor, it was the 1970 and 1972 Total Solar Eclipses in Nova Scotia which inspired Carly Simon's song "You're so Vain." There is a line in that song, "you flew your Lear jet to Nova Scotia to see a Total Eclipse of the Sun."

When the song was released, I was a twelve-year-old girl living on a farm in Central Texas. That line, to me, represented the epitome of freedom. I understood the song was about a woman complaining about an arrogant man, but I couldn't help but dream of what it would be like to *be* that arrogant man. Well, not a man, but to have that confidence, style, and wherewithal to fly a Lear jet just to see a Total Solar Eclipse, or to walk onto a yacht wearing a hat tipped over one eye and an apricot-colored scarf. The man in that song had the freedom to explore the wonders of the universe on his own terms, and I'd always wanted to know what that felt like.

Humming the song, I called American Airlines (you still had to call to make flight reservations back then) and tried to get us to La Paz, but the flights were already completely booked. With a world map and a little extra research at our local library (also no internet back then), I saw Puerto Vallarta wasn't too far from the path. I booked us for a two-week tour package.

Somewhere in my research, I'd managed to get my hands on the NASA bulletin for the 1991 Total Solar Eclipse written by Fred Espenak, the NASA Astrophysicist who eventually developed and still supports NASA's eclipse site. The bulletin contained detailed predictions, maps, and meteorology for the eclipse. Puerto Vallarta was not actually on the path of Totality, the straight-line path of the Moon's shadow on the Earth as it blocks out the light of the Sun, but it was within 40 miles. While Hawaii was a huge vacation destination to see the Total Eclipse, both Fred Espenak and Jay Anderson, eclipse experts, recommended Mexico for its better chance of sunny weather. If we could get ourselves as close as Puerto Vallarta, I had faith we could get the last 40 miles to the path to fully witness the event.

At that point in my life, I'd never been out of the country. Fortunately, this was also a time when all you needed for travel to Mexico was a driver's license as proof of US citizenship.

We arrived at the Puerto Vallarta Hilton, an older hotel that had air conditioning in the rooms, but not in the hallways or in many of the common areas. With a Mexican ancestry and a Texas upbringing, I was comfortable, but my husband was very unhappy with the conditions. July was also a difficult month for him to be away from his business. He ended up making several calls back to the states to handle business issues. We had to pay outrageous hotel-controlled long-distance rates, with each call requiring operator assistance, adding extra frustration. These calls could run $2-$5/minute, a far cry from just logging into hotel wifi and calling back to the states on Skype, Zoom, or Teams!

The hotel concierge helped us find a tour operator taking other tourists to see the eclipse from a beach within range of the Totality. We left the hotel at 7 a.m. and drove for about two hours, arriving at the beach around 9 a.m. The tour van was also not air conditioned, so it was a long, hot ride. Thankfully, the weather was moderate for July. Everything was done on paper maps back then, so I didn't keep good records of where we went exactly. These days, I keep much better recordings and record my locations on the eclipse-chaser-log.com website.

The beach was beautiful with a quaint beach bar/store/chair rental operation. We got a couple of chairs and set ourselves up alongside dozens of other eclipse tourists with cameras and glasses. Several more vans drove up and dislodged more tourists as the first contact time grew closer. This is referred to as C1 (Contact 1) and is the moment when the Moon takes its first tiny nibble out of the solar disk.

I had several point-and-shoot cameras and one high-end camera I borrowed from a friend with me to record the event. Like many first-timers, I didn't understand just how strong the Sun was and made a rookie mistake. I set up the cameras facing the Sun without putting filters on first. It didn't take long for the Sun's rays to burn out the backs of the cameras. None of the pictures turned out, whether they were supposed to be of the eclipse or of other things we did on that trip. I ended up buying my friend a replacement camera once I realized I'd damaged hers beyond repair.

With eclipse glasses (safety first!) I'd purchased prior to the trip. I watched as the Moon took its first bite of the Sun. The light gained a strange purplish cast as the Moon continued its procession. When the crescent covered about 80% of the Sun, I noticed the sand looked extra sparkly. Little rainbows caught at the corner of my eye. An experienced eclipse chaser pointed out the shadows coming between the leaves of a tree, creating mini crescents.

I settled on the sand watching the sky through my eclipse shades. When totality started, someone yelled, "Glasses OFF!" I removed my shades and stood up in awe. The Sun was *gone*. In its place, a beautiful corona shot streamers from behind the black hole of the Moon. Even though I knew scientifically what was happening before my eyes, a small pit of primal fear opened in my stomach.

A slight breeze came up as Totality hit. I became profoundly aware of the darkness of the sky beyond the corona and the surreal beauty of the sunset out in the ocean where the shadow ended. My soul resonated with the immense beauty and gift of the universe. I was both the luckiest, most blessed person in the world and just an insignificant ant on the face of the planet. For this moment in time to happen, for me to have this experience, the Sun, the Moon, the Earth, and I had to co-exist at that exact point in space and time.

When the partial eclipse (earlier stages of an eclipse prior to Totality) started, the sky was clear, but as the temperature dropped, small wisps of clouds started to form. I now know this is eclipse weather that sometimes forms, especially along coastlines. As the amazing 6 minutes, 53 seconds ticked by, the longest eclipse of the century, the cloud wisps moved in, making the end of the eclipse slightly misty. All too briefly it was over. Everyone along that beach was yelling and screaming with happiness.

That's when I asked the question that would inspire the rest of my life, "When and where is the next one?"

CHAPTER TWO

SUBSEQUENT EXPERIENCES ARE NO LESS AWESOME

To quickly let you know a little about me, I am Leticia Ferrer, an umbraphile, a person who travels planet Earth to stand in the path of the Moon's shadow and be awestruck by Total Solar Eclipses. As of this writing, I've experienced 18 Total Solar Eclipses, two Annular, and one partial. According to the eclipse-chaser-log.com site I mentioned earlier, I'm among the top 30 people sorted by number of total eclipses seen in a lifetime. I've traveled all over the world, visiting six continents and a fly over of the seventh, Antarctica. My friends consider me either inspired or crazy. I can never retire because I'll always be working to pay for the next eclipse. My personal mission in life is to see every Total Solar Eclipse on the planet until the day comes that I don't have the health to make the wealth to see the next one. For now, I plan to live to 104, which will get me 50 Total Solar Eclipses in my lifetime.

Regretfully, in 1992, I had not yet embraced the concept of world travel to chase eclipses. I had the desire to see more, but I let the distances stand in my way. I learned the next four eclipses were:

- June 30, 1992 making landfall only in Uruguay, South America
- November 3, 1994 crossing the midline of South America through Peru, Chile, Bolivia, Paraguay, Argentina, and Brazil
- October 24 1995 crossing over Iran, Afghanistan, Pakistan, and India, Burma, Thailand, Cambodia, and Indonesia
- March 9, 1997 Northern Mongolia and Russia

Has that ever happened to you? You want to do something, visit somewhere, take on a new challenge, but you stop yourself. You limit yourself by putting up your own barriers, preventing yourself from pursuing your dreams. You fail to step through an open door of opportunity. That is what I did. I thought it was too expensive or too scary to go to some of those places. My world stayed within the boundaries of North America.

I also had other hopes and dreams at the time. I was 31, and we had plans for children. The only sadness in my life is that children never happened for us.

Although we didn't go to those four eclipses, I heard stories that fueled my dream of seeing another one. Happily, the opportunity came in 1998 with a cruise in the Caribbean. I booked a cabin on the Monarch of the Seas, a ship of the Royal Caribbean line scheduled to be on centerline (the point where Totality lasts the longest during an eclipse) in the waters near Aruba at the time of the eclipse.

An eclipse at sea is amazing. There were about six ships positioned within sight of the Monarch of the Seas for the viewing.

The day of the eclipse, the deck was filled with umbraphiles (people who chase eclipses), large cameras, and telescopes. It was a beautiful, sunny day on the blue waters, and we had high hopes.

I remember being so excited sitting on a deck chair preparing for the event. Again, the sun disappeared and again, I felt the way I did on that Mexican beach eight years previous. I sat enthralled as the expanse of the corona surrounded the black disc of the Moon and the desire to see another one filled my heart almost before it was over. But this time, something changed for me. What changed was the opportunity to meet and get to know several of my fellow umbraphiles. Being on the cruise gave me a chance to learn from them.

At dinner, I sat with people who'd seen five, six, sometimes more than 10 Total Solar Eclipses. We'd sit with them in the bar having drinks, listening to their stories of traveling to India, Mongolia, and other places in search of the Moon's shadow. Listening to them expanded my world. Those boundaries I'd perceived dropped away. It was on that cruise that I decided to see every Total Solar Eclipse on the planet for the rest of my life.

As if to punctuate my decision, a volcano on Montserrat erupted just after the eclipse as we sailed by.

Today, I am a person who has walked on six of the seven continents and experienced 18 events in Asia, South America, North America, Africa, Europe, and Australia. This winter, I'll see the December 4, 2021 Total Solar Eclipse in Antarctica and finally step on the Antarctic continent.

For the eclipse in 1999, we went on another cruise, this time in Europe. Our celebrity host was the former astronaut Buzz Aldrin. The best part of the trip for my husband was having a drink with Buzz in the lounge. My friend Kathy Connolly joined us on that trip. I had traveled with her to Greece in 1998 for a week. The eclipse path was over the North Atlantic and it was the cruise's maiden voyage. The morning started cloudy, then got worse. Kathy and I stoically stayed on deck with other hopefuls as the captain of the ship, with Buzz at his side, navigated us through a squall looking for a bit of clear sky. Totality started just as the captain successfully positioned us to get a "lucky hole" in the clouds.

The next eclipse was on June 21, 2001, crossing the African countries of Angola, Zambia, and Mozambique with over three minutes of Totality.

Africa. The thought of traveling there brought up thoughts of adventure, mystery, and a bit of fear. I decided to do an Eclipse Expedition Tour with a company called Sita Tours, an experienced tour provider for Africa and Asia. Due to his limited mobility, my husband decided not to join me.

Kathy came instead. She would travel separately and meet me at the Intercontinental Hotel in Luska, Zambia.

This was one of my more adventurous trips.

I'd been doing consulting work for a few years by then and amassed quite a few travel points which I turned in for first-class tickets on Swiss Air to Johannesburg. This was an older airline that had not yet upgraded to lay down seats. Still, it was a comfortable flight.

Standing in passport control at the Johannesburg Airport, the line at least 45 minutes long, I looked up to see men dressed in camo with submachine guns patrolling above us. The line of fire into the passport area was intentionally kept clear. It made me uncomfortable watching these men walking on the balcony above us holding guns at ready.

At the same time, a strong feeling of pride and (I'll admit) smugness came over me. As an American, I felt so privileged and lucky to live in a country that didn't need to do this at our airports. My smugness was shattered only a few months later on September 11, 2001.

The World Trade Center Attack changed air travel forever. Sadly, we joined the rest of the world in the heightened vigilance against global terrorism. I live 20 minutes from the airport. I used to leave my house one hour before flight time. It took less than 10 minutes to park and zip through security. I'd walk in with my luggage, check in just as the plane started boarding, then walk on to the plane just in time. I miss those days of blissful travel.

When I went to the airport in October 2001 on another trip, I remembered my moment of smugness in South Africa as I politely nodded to the camo-clad young solider holding a submachine gun at the ready as I passed through security, having left my house three hours before flight time to be sure I got through on time.

In Africa, I ran into a little trouble after I went through passport control and gathered my luggage. I went to the ticket counter to get checked in for my next flight to Lusaka, Zambia. I had purchased my connecting tickets from Swiss Air.

But the ticket desk said I was not booked on the flight to Luska and all the direct flights were full. Of course, they were. Everyone was coming in for the eclipse. This was during the transition from paper to online ticketing for most airlines. Luckily, Swiss Air had not made the transition yet, and I had my paper ticket proof in hand.

Swiss Air put me on a plane to Harare, Zimbabwe, with a connecting flight to Lusaka. However, I'd have to stay overnight in Harare. I took a deep breath just now as I'm writing this remembering how I felt, that fear of going into a travel situation without a plan, someplace potentially unsafe. At least I was still first class. Once on the plane, I engaged with my seat mate, an experienced traveler, and asked for the best hotel to stay in Harare. He recommended the Meikles hotel in central Harare.

An 'approved' taxi from the airport delivered me safely to my hotel. My room overlooked the Africa Unity Square Park and Market.

Still wired from the time zone change, I asked the front desk about the market. It catered to tourists and was considered safe. I walked across the street and started doing some shopping. A local who knew English attached himself to me, but not seeing anything I really wanted to carry with me for the three weeks of travel, I disappointed the stall owners.

With the market closing, and I being the last one there, they became more desperate to sell me something, becoming more aggressive and starting to mob me. Rather than cowering under the mob, I pulled out my inner Texan and told them to Back the F... off. Everyone calmed down. I purchased a small giraffe and elephant and left.

That moment brings up strong feelings for me to this day. By the grace of God, those sellers could be me in Mexico. My grandfather left Mexico at the age of 17, crossing the border, legally at the time in 1917 by paying a 'dime'. That act and his choice to raise his family in Texas gave me, his granddaughter, opportunities that I've taken/earned to become educated and working in this country.

My air travel adventure continued the next day when my 'puddle' jumper to Luska failed to show up after eight hours. By this time, I had run into several other eclipse travelers with the same destination. We ended up on another airline and arrived in Lusaka only half a day late to join the tour.

Although I'd met other umbraphiles on my cruise, it was on this tour that I started accumulating life-long eclipse friends. I met Lynn Anderson on our first excursion on the SITA tour around Lusaka. At the time, Lynn was a school teacher in Sonoma County, California, and an avid umbraphile. I've since either toured with Lynn or met up with him on our various eclipse travels. Sadly, he passed this year. I'll miss his company.

Oliver Staiger, or "Klipsi, Paparazzo Del Cielo" as he calls himself, was also on that trip. His website is klipsi.com. When I met him, he was a chauffeur in Switzerland and a devoted eclipse chaser. Later, Oliver became a tornado chaser, and we visited when he flew through Dallas on his way to Oklahoma to chase tornados. These days he is leading tours to Iceland for the aurora borealis. He considers his primary occupation as "artist."

My most inspirational eclipse traveler friends were on this trip as well. Two retired Wesleyan professors, one who taught genetics, the other chemistry. During the tour, I had several dinners together with these engaging, well-traveled women. Every time I sat down with them, it was another story of travel to Russia in the 70's and 80's before the fall of the wall.

Or of travel to China before Walmart opened up it up in the 90's.

I remember saying to them, "I want to be you when I grow up." I have a vivid memory of a balloon trip with them over Victoria Falls. One of the ladies used a walker to get up to the balloon lift off station. Nothing stopped these ladies from enjoying life.

Sita Tours arranged for us to experience the eclipse from a local farm located on centerline. The location was about a one-hour drive over rough roads from where we were staying. Once we got there, though, it looked surprisingly like any farm back in Texas. Everyone set up in one of the fields. I and others had brought a sheet to lay out on the ground to capture shadow bands.

Most people don't know what shadow bands are. These tricks of the light appear on the ground or sometimes on the walls of buildings just as totality starts or ends. They are thin, wavy lines of alternating light and dark that can be seen moving and undulating in parallel on plain-colored surfaces immediately before and after a total solar eclipse. They resemble the light shadows you see at the bottom of a pool.

The eclipse was high in the sky at 2:11 p.m. local time and lasted 3 minutes 16 seconds. It was a nice day with the temperature in the mid 70°F's. As the moon moved to second contact, the point at which a Total or Annular phase of the eclipse begins, the sky became a beautiful, brilliant blue.

I remember looking around just before second contact, at all those around me, the wonderful sky, looking through the filters as the Moon took a bigger and bigger bite out of the Sun, and just being at peace with being there, in the middle of Africa, preparing to see something wonderful. I felt an immense gratitude for my life, my existence.

Just before the eclipse started, we were rewarded with shadow bands flowing across the sheets we'd spread on the ground. When first contact started, we noticed a lot of sun-spot activity as the Sun was near its maximum 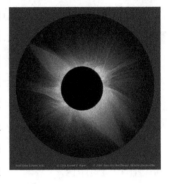 in its 11-year solar cycle. The solar cycle, or solar magnetic activity cycle, is a nearly periodic 11-year change in the Sun's activity measured in terms of variations in the number of observed sunspots on the solar surface. Levels of solar radiation and ejection of solar material, the number and size of sunspots, solar flares, and coronal loops are at a maximum at this time, creating a more extensive corona during an eclipse. This eclipse did not disappoint! The corona felt alive with flares and wings. The 2024 Total Solar Eclipse is also close to Solar Maximum, so I'm expecting a wonderful, expansive corona.

We celebrated that night with a traditional dance presentation and feast in a tent. Sadly, we were reminded of the poverty around us when one of my fellow tourist's cameras was stolen from next to him by someone reaching under the tent. It was a real tragedy because all his eclipse pictures were lost.

Our adventure inadvertently included a taste of the tension between Zimbabwe and Zambia. Our tour driver refused to take us across the bridge to Zambia to our hotel. He stopped about a mile away saying he did not have the correct papers to drive in Zambia. The hotel in Zambia could/would not send a bus to pick us up across the border either. Our guide did her best to get someone to cart us and our luggage across, but we gave up at about half an hour before sundown and walked the last mile toting our own luggage, some of the members still carrying all their telescope gear. We wanted to be in the hotel before dark. From the bridge, as we crossed over, we saw the local hippos happily settling down for sleep. Not something I would have seen in Texas for sure!

Part of the tour included a stay at a 'waterhole' hotel, from which we'd take safaris to see the local wildlife. Flying another puddle jumper into this part of the African savannah, I again was taken by how much this looked like Texas. Scrubby trees, greenish-gold grass, hot, dry, and a long horizon.

The hotel was elegant with dinners featuring local fare. In the middle of the night, elephants came through the property, and the staff had to chase them away, loudly. It was cold at night as this was winter in Africa, with temperatures in the mid-40°Fs.

During our safari, we got dangerously close to a lion when I leaned too far out of the jeep. We visited a local area where elephants were known to frequent.

We watched them eating the leaves off the tops of the scrubby trees. At one point, we quietly walked around a junior elephant to avoid riling his mother. And I answered a call of nature using the local currency for TP. Quite the adventure.

Staying up late with some of the astronomers of the group, I saw the Southern Cross for the first time thanks to the crystal-clear skies you can only get in an area without major light pollution, though we had also asked the hotel to turn off the security lights around the property. We could not set up very far from the hotel as nights beyond the property line were unsafe.

We saw an entire pride of lions hunting one night. We were on our way back to the lodge when the guide got a call that the pride had been spotted, so back out we went. I still remember the feeling of tension in the air with the pride of lions' eyes peering out of the foliage not even 10 feet from us. As I went to bed that night, safely tucked in my hotel bed, I learned a lion's roar in the middle of a dark night will send goose bumps of primal fear up your spine no matter how solid the walls are around you.

The next day, we had another flight delay following our safari adventure. When we got to the local airport, the flight was delayed for a couple of hours by a lion sleeping under the wing.

One of my travel habits is to bring back at least some of the local currency of my travels and put it in a large flower vase I keep in my hallway. It's a daily reminder of my travels. Like most third world countries, the exchange rate in the hotel and 'on the street' differed greatly. Several of us decided to go ahead with a street exchange. It was risky and if you are planning to do this on a trip in any country, I recommend doing it with a group of three or more. We did the exchange, the money passed locally, and I had a lot of cheap souvenirs to share when I got back home. Especially since the currency further declined within a year after we got back, and the country ended up issuing a new currency. The bills I still have from that trip are now truly nothing but decoration.

There were a few more eclipse experiences after that which I will perhaps write about and place in another collection, but the next one I want to discuss here is the one on August 21, 2017, when many in the US saw the Total Solar Eclipse that crossed the country from sea to sea. I watched with 22 friends and family members from a lovely KOA campground just south of Madras near the centerline.

My brother-in-law, Brian, a Hollywood writer, thought it was a lot of hype until he saw it. Afterwards, touched and moved by the experience, he looked at me and asked the classic question, when is the next one?

Brian joined my husband and I on the next eclipse in La Serena, Chile, on July 2, 2019. On that trip, I met about 20 second time eclipse chasers, whose first eclipse was the 2017 event in the USA. For some, standing under the Moon's shadow, seeing the Sun "eaten" so she can display her beautiful corona, prominences, bailey's beads, and diamond rings is/are life changing moment(s). Talk with one of the umbraphiles, those who "chase" Total Solar Eclipses, we can tell you all about it. However, it is only when YOU stand in the shadow of the Moon yourself, aligned with the Sun and Moon in the right position on Earth on a clear day, that you can understand what motivates us to travel the world, far from our homes, to be there in the shadow again and again.

I'm especially excited about this upcoming April 8, 2024 Total Solar Eclipse in Texas as I live in Farmers Branch, Texas and will get 3 minutes 19 seconds at my home. My parents' farm near Corsicana is only 10 miles from centerline and will get 4 minutes 19 seconds. This will be in April, so much of the decision on where to view from will depend on the local weather as we get nearer to the eclipse date.

CHAPTER THREE

WHY FOCUS ON TEXAS?

You've seen that the focus of this book is Texas, Texans, and the history of other Total Solar Eclipses in Texas.

Within a six-month period of time, the State of Texas is going to be lucky enough to host the October 14, 2023 Annular and the April 8, 2024 Total Solar eclipses. These are the first Annular and Total Solar eclipses crossing the USA since August 21, 2017. For Texas, this will be the first Total Solar Eclipse crossing the state since the famous July 29, 1878 eclipse. In fact, this will be only the second Total Solar Eclipse crossing the State of Texas in the state's history. Texas was a Spanish Colony when the previous Total Solar Eclipse of 1778 grazed what is now the Texas Coast.

Many will likely travel to other states or perhaps even farther south to Mexico since the weather in Texas will be 50/50 for either time of year, but as Texas is my home state, I'm excited to view it from as close to home as makes sense.

Those looking for the best weather prospects in the USA or perhaps even to view from the "cross path" of both eclipses will likely head to one of the prettiest areas of Texas, the Hill Country between San Antonio and Austin. The Texas Hill Country is blessed with beautiful rolling hills with several rivers including the Guadalupe, South Llano, and Frio. These rivers cut through the hills creating wonderful views from natural escarpments created by the rivers over hundreds of years. Ten of Texas's state parks are in the path of both events (see the chapter Upcoming Texas Eclipses for a chart of Texas parks to visit for these events).

The lovely city of San Antonio, with its historic past and its world-renowned 15-mile urban waterway The River Walk, is sure to be a favorite for eclipse travelers. The River Walk, tucked quietly below street level and only steps from the Alamo, is the city's treasure.

Even better, the city is fully in the path of the October 14, 2023 Annular eclipse. Some of its northwest areas will also be in the path of the April 8, 2024 Total Solar Eclipse.

Austin, the state capitol, is outside the path of Annularity on October 14, 2023, but the northwest areas of the city will see Totality on April 8, 2024.

As a proud Texan, my home for the last 50 years, my hope is this book will do two things:

First, introduce Texans to the upcoming Annular and Total Solar Eclipses. As of 2020, the population of Texas is about 30 million, with three of our largest metroplexes — San Antonio, Austin, and Dallas/Fort Worth — in the path of Totality on April 8, 2024, and San Antonio getting the premier opening act of the Annular eclipse on October 14, 2023. This book contains information on where to see these events, what to expect and important safety information.

Second, I want to introduce Texas to those planning to join us here in our great state for these events. With all the recent growth in Texas from companies moving their headquarters here, it's likely you have friends and/or relatives willing to welcome you to share this event with them. We are a proud state, with lovely areas to visit, both in and outside of the eclipse paths. From the mountains around Big Bend and Fort Davis where our world-renowned McDonald Observatory resides, to the Texas Hill Country, the pastureland of North Texas, and out to the Big Thicket areas of East Texas, we have more than 80 state parks, many with camping facilities to bring your friends and family with you to enjoy our great state. Travel to and within Texas is easy with our great Interstate/Texas Highway systems, and easily accessible airports, or you can recreate the travel of the 1878 eclipse by arriving by train.

My hope is that at least three million people, Texans and others, get to experience and enjoy the Total Solar Eclipse and the Annular Solar Eclipse from our great state.

As a lot of eclipses are hard to get to for many in the United States, their first or second eclipse will be the 2024 Total Solar Eclipse crossing Texas, Oklahoma, Missouri (again), Illinois (again), Kentucky (again), Indiana, Ohio, Pennsylvania, New York, Vermont, New Hampshire, and Maine. The 2024 eclipse also crosses Mexico (the first since 1991) and Canada. The last eclipse to cross any part of Canada was the August 1, 2008 Total Solar Eclipse, which started at sunrise in the Canadian Artic Circle.

On April 8, 2024, Texans in a 120-mile path from Mexico, through Austin, Waco, Ft. Worth, Dallas, and Texarkana will see the spectacular sight of a Total Solar Eclipse. All you need to see it is to get in the path!

Finding the path has been made easier through the work of Xavier M. Jubier, a prominent umbraphile. He created interactive maps using Google Maps and the Five Millennium (-1999 to +3000) Canon of Solar Eclipses Database created by Fred Espenak (-mreclipse.com) and Jean Meeus. To show you how helpful they are, I'll be using screenshots from Mr. Jubier's maps throughout the rest of this book. If you find them as helpful as I do, please click on the "donate" button on the site. Small donations help support the cost of all the traffic this site gets during such popular eclipses.

The path of the 2024 eclipse is the section in this map shaded between the two red lines. The blue line between the two red lines is centerline, the point of maximum Total Solar Eclipse along the path.

The closer you are to the blue centerline, the longer Totality you'll see. To see if your house in Texas will be in the path, check the map.

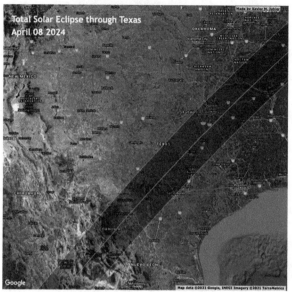

http://xjubier.free.fr/en/site_pages/solar_eclipses/ TSE_2024_GoogleMapFull.html

Following is a list of the key cities in Texas that will be in the path of Totality on April 8, 2024. Cities bolded will be in the path of the October 14, 2023 Annular Eclipse also.

As I mentioned above, the April 8, 2024 Total Solar Eclipse crosses a large swath from Mexico, up the USA starting in Texas, and on through to

Total Solar Eclipse of April 8, 2024 Approximate Duration	
City	Duration of Total Eclipse
Del Rio	3m 22.5s
Eagle Pass	4m 23.9s
Uvalde	4m 15.6s
Kerrville	4m 24.6s
Junction	3m 07.0s
Fredericksburg	4m 24.1s
Austin	1m 44.2s
Killeen	4m 17.1s
Waco	4m 11.9s
Hillsboro	4m 22.4s
Ennis	4m 22.5s
Dallas	3m 50.7s
Ft Worth	2m 29.9s
Tyler	1m 52.3s
Texarkana	2m 25.0s

Canada, as shown in the map below courtesy of eclipse-maps.com.

Credit Mike Zieler -
◻GreatAmericanEclipse.com

| Annular Solar Eclipse of October 14, 2023 (Times CDT) | |
City	Duration of Total Eclipse
Midland	4m 54.9s
Odessa	4m 49.0s
San Angelo	3m 12.9s
Fredericksburg	2m 14.5s
Junction	4m 30.0s
Kerville	4m 16.2s
Uvalde	3m 40.4s
San Antonio	4m 23.2s
Corpus Christi	5m 02.2s

Leading up to this event is an Annular Eclipse on October 14, 2023. During an Annular Eclipse, the edge of the Sun remains visible as a bright ring around the Moon. This event will cross over southwestern Texas through Midland/ Odessa, San Antonio, and Corpus Christi. The cities bolded in this table will also see the April 8, 2024 Total Solar Eclipse. To see if your house in Texas will be in the path, check the map.

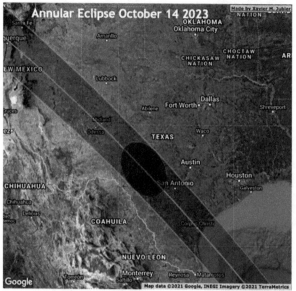

*http://xjubier.free.fr/en/site_pages/solar_eclipses/
ASE_2023_GoogleMapFull.html*

Some locations in Texas will be in both paths. For example, the section of US Interstate 10 between San Antonio and Junction will see both eclipses as will part of Texas Hwy 90 between San Antonio and Brackettville and Texas Hwy 377 between Interstate 10 and Carta Valley.

For so many reasons, you will not want to be driving down these highways during the event. Much better to reserve yourself some comfy space in one of the many Texas state parks conveniently located along the path. See the chapter on Upcoming Texas Eclipses for information about the best parks to visit.

If you are planning to view your second, third or more eclipses, I also encourage you to log these events at eclipse-chaser-log.com. The current record stands at 35, held by Jay Pasachoff, a Field Memorial Professor of Astronomy at Williams College.

CHAPTER FOUR

THE TEXAS ECLIPSE OF 1878

T o understand the excitement of a solar eclipse, it's important to understand eclipses are rare at any one location on Earth. The mean frequency for a total eclipse of the Sun for any given point on the Earth's surface, calculated by Jean Meeus of Belgium, is once every 375 years. This is made even more rare since the path of an eclipse averages only 60-70 miles in width.

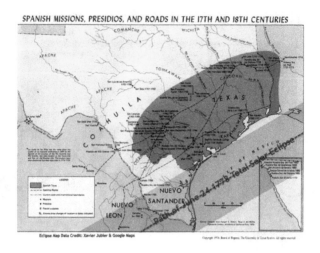

SPANISH MISSIONS, PRESIDIOS, AND ROADS IN THE 17TH AND 18TH CENTURIES

Eclipse Map Data Credit: Xavier Jubier & Google Maps

The last time Texans experienced a Great Total Solar Eclipse was July 29, 1878. Let me rephrase that, the FIRST time TEXANS experienced a Total Solar Eclipse was the July 29, 1878 eclipse. Eclipses, both Annular and Total, are as rare in Texas as they are for anywhere. For a state as big as Texas, only one other eclipse crossed what is now Texas since the founding of the USA in 1776. That Total Solar Eclipse was on June 24, 1778. Its path crossed Mexico, went along the coast of Texas, and on through what is now Louisiana, Alabama, Georgia, South Carolina, North Carolina, and Virginia. However, this was when Texas was still a Spanish Colony. The only European settlements in the path in 1778 were Monterrey, Revilla, Meir, Carmargo, and Reynosa south of the Rio Grande in Mexico.

There were no other eclipses in the 100 years from 1778 until 1878 that crossed what is now Texas. Texas was founded in 1836, so the first Total Solar Eclipse experienced by TEXANS was July 29, 1878.

The eclipse path started in Russia, came north through Alaska (at that time owned by Russia), through Western Canada, hit the US starting in the territories (not states) of Montana and Wyoming, through the state of Colorado, the territory of New Mexico, Indian Territory (now Oklahoma), and the state of Texas, ending just before Puerto Rico in the Caribbean. There were scientific parties across the US that day studying the eclipse. Most of them were in the other states. The parties in Wyoming and Colorado are covered very well in the American Eclipse book by David Baron.

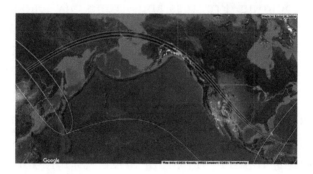

Fort Worth was centerline for this eclipse, approximately at the blue dot on this map from the *Chicago Times* July 22, 1878 edition, a fact that makes any city an attraction for those planning to study the eclipse.

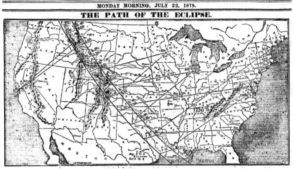

Some context about Texas and Ft Worth in 1878. The Civil War ended in 1865, and economically Texas was booming. Progress and strong city leadership made Ft. Worth an attractive option for the eclipse. Ft. Worth got its first railroad station in 1876, and the city boasted the El Paso Hotel. The El Paso Hotel was lauded as a "monument to the liberality of our public-spirited citizens." The El Paso was the city's first three-story building, a solid stone edifice fronting 100 feet on Main Street and 130 feet on Fourth Street. This luxury hotel boasted the best ventilated and gas-lighted rooms plus a spacious and elegant billiard room. Ventilation was something most structures of the time didn't have and badly needed in the Texas summers. A reporter for the July 21, 1878 *Ft. Worth Daily Democrat* interviewed Leonard Waldo of Howard University in St. Louis who led an expedition to Ft. Worth to see the eclipse in his apartments at the El Paso. More about Leonard Waldo and his expedition later.

In 1878, Ft. Worth had reached a population of about 7,000 people, having just incorporated about five years earlier. The economy centered around cattle as they moved up the Chisholm Trail to the Kansas Railhead. Though the famous stockyards were on the map as it were, they had not been built yet. Those were established in 1889. The mayor was R.E. Beckham, and the marshal was a colorful man named Jim Courtright. Streets were unpaved, parking was a hitching post, and no one had piped water or indoor plumbing. Ft. Worth was still the Wild West, where guns were worn as a necessity for defense not only from other men but from the common rattler and other critters. Only a few years before, Ft. Worth, in the middle of an economic downturn, had a myth spring up that a panther was sleeping in the street undisturbed. While the comment was meant as an insult, Ft. Worth took it on as a mascot, leading to the local school's Panther theme.

It was a small town with grand ambitions. It had nine churches, nine schools, two militia units and a dozen physicians. With the economic boom based around cattle and railroad, the town also hosted the infamous "Hell's Half-Acre," a designated "Red-Light District." The half-acre block was originally designated from Tenth to Fifteenth streets while intersecting with Houston, Main, and Rusk with Throckmorton and Calhoun streets established as boundaries.

Railroads were expanding and booming throughout the nation, making travel across the US more available to the less pioneering and adventurous. Railroads boasted luxurious cars and coaches. Travel time was reduced from weeks or even months in wagons, stagecoaches, or horses, to a relatively smooth trip undertaken in a matter of days. The railroads were spreading throughout Texas, too, connecting Texas cities to the rest of the US. The newly formed Texas and Pacific Railway connected 24 cities in Texas, including Ft. Worth in 1876, to Chicago, St. Louis, and Memphis. This ease of travel from established modern cities to the outskirts of settlement brought an edge of civility to the wild west.

However, the railroad to Ft. Worth almost didn't happen in time for Waldo to travel to Ft. Worth in July of 1878.

The campaign to bring a railroad to town began in 1858, led by Ft. Worth citizens. However, it wasn't until 1872 that the Texas & Pacific Railroad offered assurance that the town would be on the T&P route as track was laid westward across Texas from Longview toward the Pacific Ocean. The Texas legislature promised T&P a land grant of 10,240 acres of land per mile of track and Ft. Worth threw in another 320 acres for the passenger depot, roundhouse, and railyard a mile south of the courthouse. The only catch was it had to be finished by January 1, 1873. When T&P missed that January 1, 1873 deadline, the legislature and Ft. Worth extended the deadline.

Even with the missed deadline, the mere promise of a railroad caused a boom in Fort Worth and spurred the city to officially incorporate on March 1, 1873. By August 1873, the T&P railroad had tracks from Longview to Dallas. Fort Worth was just thirty miles away. One of Ft. Worth's leaders and the editor of the *Ft. Worth Daily Democrat* newspaper drew a now famous "tarantula map" that showed Fort Worth as a bona-fide railroad hub in the near future.

Just as things were moving along, Black Friday, a stock market crash, hit on September 18, 1873, triggered by the closure of Jay Cooke and Company, a banking firm heavily invested in railroad construction. Back then, investment in railroads was like investment in the dot coms in the late 1990's and early 2000's. Eighty-nine of the country's 364 railroads crashed into bankruptcy and unemployment reached 14%. The T&P track stalled at Dallas and Fort Worth's population dropped to 600.

But T&P persisted and by 1874, the T&P got the track six more miles west to Eagle Ford, just 24 miles from Fort Worth.

By then, delegates to the state Constitutional Convention of 1875 stipulated that T&P would lose its promised land grant if Fort Worth was not reached by the adjournment of the legislature, set to meet under the new constitution of 1876.

Faced with this new deadline, Fort Worth pitched in and created a construction company to put down the roadbed and bridges. The work continued to lag to the last minute. Down in Austin, the state legislature had finished its work and wanted to adjourn. T&P was going to lose its land grant. The only person in the way was Tarrant County's representative in the Texas House, Nicholas Henry Darnell. In ill health as the House began to vote to adjourn, Darnell was carried into the House chamber on a cot for 15 days to vote "nay" to adjournment in the interest of giving Ft. Worth, the primary city in his jurisdiction, a chance.

Desperate to make things happen, the citizens of Ft. Worth were driving out to the track to help with the work. Finally, at 11:23 a.m. July 19, 1876, Fort Worth's first train came into town with a shrill scream. Down in Austin, Tarrant County's Representative Nicholas Henry Darnell could vote "yea" and go home to get well.

The 1878 Ft Worth Eclipse Party

Thus, the track was laid for Mr. Leonard Waldo and his "Ft. Worth Eclipse Party" to make their way to the eclipse zone via railroad in mid-July 1878. The trip from St. Louis to Ft. Worth was only two days by rail rather than weeks by buggy. This larger, more reliable mode of transportation also allowed massive scientific equipment to be shipped with less risk of breakage.

*Tarrant County College NE Heritage Room in
partnership with The Portal to Texas History,
a digital repository hosted by the University of
North Texas Libraries*

Suddenly, scientists could bring what they needed
to study the eclipse such as sensitive telescopes and
complex photographic equipment.

To study this historic eclipse taking place so close
to their home campus, the Washington University of
St. Louis sponsored the five-man team of academics
to travel. The expedition members were:

- Mr. Leonard Waldo, of Harvard College
Observatory
- Mr. R.W. Willson, of Harvard College
- Professor J.K. Rees, of Washington University,
St. Louis
- Mr. W.H. Pulsifer, of St. Louis
- Mr. F.E. Seagrave, of Providence, R.I.

There are several factors that make a
location attractive to scientists, astronomers, and

umbraphiles that are as true today as they were in 1878:

- **Location**: In the path as close to centerline or max eclipse point. Ft. Worth in 1878 was centerline. During July 29, 1878, the Max Eclipse Duration in British Columbia, Canada was 3 minutes 11 seconds, but by being on centerline, Ft. Worth got 2 minutes and 33 seconds of Totality. For the 2024 Eclipse, Ft Worth will be in the path, but not as close to centerline. The Max Duration is 4 minutes 23 seconds in Durango, Mexico. For the April 8, 2024 Total Eclipse, the location at Adams Street and Vickery, where I estimate as the earlier expedition site, will experience an estimated 2 minutes and 33 seconds of Totality, just one second off from the eclipse of July 29, 1878.
- **Weather** at time of the eclipse: Weather is the nemesis of eclipse chasers in any age. In my eclipse chasing travels, I've visited a disproportionate number of deserts and oceans trying to maximize "Clear Skies." Clear skies are the norm in Ft. Worth in July, except for the occasional summer thunderstorm. It's just hot.
- **Accessibility**: This is what made Ft. Worth attractive in 1878 as an expedition point. Other Texas cities were also attractive such as Corsicana and Waco. However, Ft. Worth had the train which made travel there so much easier and all the equipment for such an expedition could be brought with less hassle. This was quite attractive to comfort-loving academics.

- **Access to support**: The 1878 expedition leader Leonard Waldo had a letter of introduction to Spotswood Lomax, a Ft. Worth bank cashier. Further research by Mike Nichols, whose website hometownbyhandlebar.com provided much of this material, noted that Spotswood was originally from St. Louis, likely Mr. Waldo's introduction resulted from inquiries around, "Who do you know in Ft Worth, TX?".

- *Report of the Observations of the Total Solar Eclipse, July 29, 1878, Made at Fort Worth, Texas*, I estimate Mr. Lomax's Farm to be what is now the corner of S Adams Street and Vickery. If you are a historian and read Mr. Waldo's report and came up with a different location, please let me know. I'd love to hear how you determined the location.

- **Access to communications**: Ft. Worth had a telegraph office. In reading Mr. Waldo's report, I was surprised at the number of pages dedicated to determining location and time of day by communicating by telegraph back to the Washington Naval Observatory, which set the standard time in those days. I'll write a bit more on this later.

The five-man expedition arrived in Ft. Worth on July 18, 1878, loaded with boxes of equipment and a plan to study and capture the Total Solar Eclipse experience in scientific terms. The expedition hired horses and mules to carry the equipment to

Spotswood's Farm about one mile south of the Texas and Pacific Railway Depot.

As I've suggested already, the exact location of the first expedition has been lost to time. From what I gather from the report, it was on Adams Street about a mile southwest of the train depot, hence my estimate the location was near what is today South Adams and Vickery. A marker was to be placed at the courthouse to denote the location of the eclipse party, but sadly, it either was never installed or lost to history, possibly when the courthouse was rebuilt in 1881, or rebuilt again in 1895.

Mr. Waldo recruited help from the local population as well. Local residents were requested to drop by the El Paso hotel to "set their watches" to a watch that had been set to the local time and synchronized with the time at the Naval Observatory in Washington D.C.

Chester A. Arthur in 1884. Per NAD 83 (North American Datum of 1983), the offset to the prime meridian is 77°3'5.194"W for the old Naval Observatory Dome.

Waldo's report puts their location, as judged by the observation of the stars by all five members of the expedition, with their differences averaged at:

N. Latitude, 32° 45' 19"

W. Longitude, $1^h 21^m 7^s$-57

I'll leave it to those familiar with sideral calculations to make the translation on the Longitude from the Naval Observatory Meridian to the Prime Meridian. My estimate is that they were at

32.743352N and 97.3358.63W per Google Maps.

Time and Space in 1878

Just a few thoughts as to the calculations of Latitude and Longitude for this expedition. Of Waldo's 55-page report, 16 pages or 29% of the report was dedicated to determining their location in relation to and trying to synchronize watches back to the Naval Observatory in Washington D.C.

In 1878, all watches were of the wind-up variety and lost time regularly. In addition, there were no time zones like we have today. All timepieces (clocks and watches) were set to the local noon of the nearest courthouse across the country. Waldo and party spent several nights trying to get a clear telegraph line back to the Naval Observatory. They were unsuccessful and ended up synchronizing with David Todd, another professor basing his observations out of Dallas, who had been able to synchronize with the Naval Observatory.

Everyone in the party were dedicated to determining their location by observation of the stars. While they all followed the same program to determine the party's location, they came up with slightly different results. The effort was made even more difficult by: "The extreme heat, the annoyance caused by the numerous insects which fluttered into the observers' faces, or crept under their clothing, or got into positions such that a movement of an instrument would jam them under pivots or under a level in reversing."

The expedition set a transit pier by digging a trench three feet deep. Transit piers provided a foundation for much of the heavy scientific equipment in the unstable environment of study in the field. The large telescopes that travelled with the party required a stable foundation to limit shaking of the pictures. While photography was available, people still relied more on drawing to record the eclipses. Photography at the time was a complex process that required dark boxes and expertise in processing the plates. Large physical exposure plates had to be switched out for every new frame or image. Moving pictures, film, was yet to be developed. The first moving picture – the Kinetograph was still 10 years away. To catch images in quick succession, the only option was to set up multiple camera boxes. Photography in the field was a complex and challenging process, a far cry from where we are today with our high-quality, multi-pixel phone cameras. Sketches competed with and often were better than photographs at the time. Take a look at the comparison between the best photograph and the best sketch from the Ft. Worth Eclipse Party as an example.

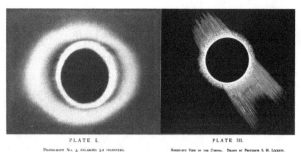

PLATE I.

Photograph No. 3, enlarged 3.2 diameters.

PLATE III.

Naked-eye View of the Corona. Drawn by Professor S. H. Lockett.

Report of Observations made of the Total Solar Eclipse , July 29, 1876, made at Fort Worth, Texas, ed. Leonard Waldo, Cambridge: Press of John Wilson and Son (1879)

The human reaction to witnessing an eclipse was another mitigating factor, the same thing that happens today. Once the eclipse starts, awe and wonder sweep away reason and even the most practiced, committed scientist forgets their intentions. Waldo wrote, "Just as the time-keeper called two, however, I noticed that I had forgotten to remove my shade-glass, which I now immediately did, regretting that I had lost 15 seconds of precious time."

During this eclipse, there is evidence in his journal that Waldo saw shadow bands:

> There were flitting shadows moving swiftly toward the east and south-east. They seemed to me regular in their occurrence, both as to time and distance apart; but how far apart they were or how many occurred in a second, I cannot form an estimate, other than that they seemed like dark crests to waves, and might

be a hundred feet apart, occurring three in a second. The faculae are seen clearly up to within say 2" of the Moon's limb. I think they are sharp to the very edge. The Moon's limb has something of the "boiling" appearance; and I am not sure that the Sun's photosphere is sharply defined in the jagged edges of the Moon's advancing limb. 2° There seems to be a minute point of light separated from, but in the line of, a prolongation of one of the Sun's cusps. (This was after second contact.) I do not know whether this is a phenomenon similar to that which produces "Bailey's Beads" or not.

While the amazing experience of the solar eclipse threw Waldo for a moment, he managed to capture a rather eloquent description of what he witnessed.

Just at the moment of greatest obscuration by the Moon, light clouds, that had for a few moments partially hidden the Sun, passed away, and the corona seemed to burst out from around the Moon, showing its spherical shape so perfectly that it seemed like a huge ball ; ... while above it a brilliant line of sunlight blazed out so narrow that I was enabled to remove the colored glass from before the eye-piece, and observed without experiencing any unpleasant effects to the eye. Scarcely had the corona appeared, and before any calculation had been made of its extent, there seemed to leap from the edge of the Moon, and in close proximity to each end of the line of sunlight, large prominences of a beautiful rose color. I saw two near the eastern, and two near the western ends of the line of sunlight.

Much of the historical information in this book was gathered from the website www.hometownbyhandlebar.com written by Mike Nichols, a retired writer and lover of his home town Ft. Worth, Texas.

CHAPTER FIVE

ECLIPSES ACROSS TEXAS

From the history of this eclipse in Texas in 1878, I defined Key Texas Towns as those with railroad connections listed in this advertisement for the Texas and Pacific Railway stops through Texas in 1878.

CITY OF FORT WORTH.
FOR 1878-79.

THE TEXAS & PACIFIC RAILWAY

AND ITS CONNECTIONS!

For the Most Direct and Quickest Line

—FROM—

St. Louis, Chicago & Memphis,

—TO—

JEFFERSON,	MARSHALL,	DALLAS,	SHREVEPORT,
LONGVIEW,	SHERMAN,	TERRELL,	MINNEOLA,
BONHAM,	McKINNEY,	HONEY GROVE,	PARIS,
CORSICANA,	CALVERT,	WACO,	BRENHAM,
LAWRENCE,	EAGLE FORD,	FORT WORTH,	HOUSTON,
GALVESTON,	HEARNE,	SAN ANTONIO,	AUSTIN,

And all Points in Texas.

Arrangements are now perfected by which THROUGH CARS from St. Louis via "The Iron Mountain Route," make close connection at TEXARKANA with the LIGHTNING EXPRESS Train of this Road for all principal points in the State.

This Line is Equipped with New and Elegant Coaches

—AND—

Palace Drawing Room and Sleeping Cars.

GEO. NOBLE,	JNO. W. DELANEY,	R. W. THOMPSON, Jr.,
Gen'l Supt.,	Gen'l Western Pass. Agt.,	Gen'l Pass. Agt.,
MARSHALL, TEXAS.	DALLAS, TEXAS.	MARSHALL, TEXAS.

The July 29, 1878 and the April 8, 2024 Total Solar eclipse paths cross these Eight Key Towns in Texas.

City	Total Eclipse Duration 1878	Total Eclipse Duration 2024
Corsicana	02:31.3	04:09.3
Lawrence	02:02.0	04:19.4
McKinney	01:32.1	03:02.9
Eagle Ford	02:27.2	03:38.6
Dallas	02:24.1	03:50.7
Terrell	01:57.3	04:20.9
Waco	00:59.9	04:11.8
Fort Worth	02:33.0	02:29.9

Because of the orbital dynamics of the 2024 eclipse, many towns will get a longer duration of totality than in 1878.

Ten Texas State Parks in Both Paths

Whether you live in one of these lucky cities already in the path or you are traveling to the state to see either of these events, you may be interested in staying at one of the several parks closely in line with the path for both the 2023 and 2024 eclipses. As my accounts earlier were intended to show, sometimes the experience of traveling to the site is a unique experience all on its own.

State Park	Annular Eclipse Duration	Total Eclipse Duration
Devil's Sinkhole State Natural Area	4m 55.0s	3m 30.0s
South Llano River State Park	4m 37.6s	3m 10.0s
Kickapoo Cavern State Park	5m 13.2s	3m 48.6s
Garner State Park	4m 47.7s	4m 26.2s
Lost Maples State Park	4m 59.4s	4m 25.9s
Hill Country State Natural Area	4m 56.5s	4m 04.5s
Old Tunnel State Park	3m 03.7s	4m 15.1s
Government Canyon State Natural Area	4m 36.9s	2m 58.0s
Guadalupe State Park	2m 42.1s	3m 08.8s
Honey Creek State Park	2m 48.8s	3m 06.7s

The state of Texas has 89 state parks. Ten of these are in the path of both the October 14, 2023 Annular Solar Eclipse and the April 8, 2024 Total Solar Eclipse. There are nine other state parks that are only in the Annular Solar Eclipse path and 22 other parks only in the Total Solar Eclipse path.

Just remember, solar eclipses are a big draw, so make reservations early and be flexible on location as some locations will be more popular.

All the information I'm providing here is gathered from the Texas State Park website. While I'm providing general information, you'll need to research details of your planned eclipse outing by going to https://tpwd.texas.gov/state-parks/ to make reservations and find out about any special events or restrictions in the park of your choice.

The Texas State Parks Site reservations are site specific. Depending on the park, you can reserve up to five sites for the same park and time period as much as five months in advance. The five-month reservation window opens at 8 a.m. Make your reservation early, as Annular and Total Solar eclipses will draw crowds. Also, many of the parks have vehicle limits for each site, please check with the park directly.

Day passes can be reserved for up to eight people per vehicle and up to two vehicles per arrival date.

the Texas State Parks website for the park of your choice. Fees are collected for the total cost of the reservation, including park entrance fees. Payment is accepted from Visa, MasterCard, or Discover card.

If you are planning to extend your stay in our great state or live here, the State of Texas also offers park passes for $70, a great option. You'll get a confirmation email after you make your reservation.

customer.service@tpwd.texas.gov or by phone at (512) 389-8900.

Devil's Sinkhole State Natural Area

The Devil's Sinkhole is located about three hours west of Austin and three hours northwest of San Antonio.

Access to the park is by pre-arranged tour only as it is a sensitive natural resource and home to Texas' largest Mexican free-tailed bats colony. Tours are by reservation only through the Devil's Sinkhole Society. Call (830) 683-BATS (2287) to prearrange a tour.

Local Eclipse Circumstances Devil's Sinkhole State Natural Area			
Annular Eclipse Saturday October 14, 2023		**Total Solar Eclipse Monday April 8, 2024**	
Duration of Annular Eclipse	4m 55.0s	Duration of Total Eclipse	3m 30.0s
Eclipse Start	9:21 AM	Eclipse Start	11:13 AM
Annularity Start	10:48 AM	Totality Start	12:30 PM
Annularity End	10:53 AM	Totality End	12:34 PM
Eclipse End	12:28 PM	Eclipse End	1:53 PM

Located within the Devil's Sinkhole State Natural Area is The Devil's Sinkhole. It is a deep, bell-shaped, collapsed limestone sink with cave passages extending below the regional water table. It is the best example of a collapsed sink along the Balcones Fault and within the Edwards Plateau region of Texas.

National Natural Landmark.

One of the most notable features of the park is the 50-foot-wide shaft which drops 140 feet into the cavern. The cavern itself has a diameter of over 320 feet and reaches 350 feet deep.

To protect this sensitive natural resource, the park does not allow visitors to enter the cavern.

Mexican free-tailed bats roost in the cavern from late spring through early fall. They migrate to Mexico for the colder months of the year. Each evening the bats are in attendance, they begin a nocturnal quest for food. Scientists estimate this colony consumes up to 30 tons of beetles and moths each night!

The park offers while the bats are in residence from March to mid-October. Bats are wild creatures, however, and don't always conform to the humans' tour schedule.

Check with the Devil's Sinkhole Society to see if eclipse viewing will be allowed due to the sensitivity of the natural environment to local wildlife. Another option is to stay at the nearby Kickapoo Cavern State Park and make the bat tour part of your overall stay in the area.

South Llano River State Park

This park is located about three hours west of Austin and three hours northwest of San Antonio.

Quick Facts	Number of Sites	People allowed per site	Restrooms w/showers available	Distance in	Nightly Fee
Campsites w/ electricity	58	8	Yes	Drive-in	$20
Campsites w/ water	8	8	Yes	Walk-in, 30-70 yards to sites	$15
Primitive sites (hike-in)	5	4	1.5 miles away	Hike in 1.5 miles	$30

South Llano River State Park is where the Hill Country meets West Texas. The park features a unique combination of rocky upland backcountry and a lush pecan grove river bottom.

Local Eclipse Circumstances South Llano River State Park			
Annular Eclipse Saturday October 14, 2023		**Total Solar Eclipse Monday April 8, 2024**	
Duration of Annular Eclipse	4m 37.6s	Duration of Total Eclipse	3m 10.0s
Eclipse Start	9:21 AM	Eclipse Start	11:14 AM
Annularity Start	10:48 AM	Totality Start	12:32 PM
Annularity End	10:53 AM	Totality End	12:35 PM
Eclipse End	12:28 PM	Eclipse End	1:54 PM

The South Llano River is a spring-fed river, unique among west and central Texas rivers. It has never run dry in recorded history. Two large springs, in addition to many smaller springs, supply most of the water in the South Llano.

The South Llano and North Llano rivers meet in the town of Junction. There they become the main Llano River, which flows into Lake LBJ and the Highland Lakes some 100 miles downstream.

The river is also home to the state fish of Texas, the Guadalupe bass. The Guadalupe bass puts up a great fight when anglers hook it, making it a very popular game fish. Great news is that you can try your luck at fishing and you do not need a fishing license to fish within park boundaries. Also, at headquarters, you can borrow fishing gear to use in the park.

This park provides refuge for wildlife and people on the southwestern edge of the Hill Country. If you plan to visit, bring your hiking boots, water toys, camping gear, and sense of adventure. Because the river is spring-fed and slow-moving, it is great for family water fun. The area is popular for families to swim, float, paddle, and fish. There are two miles of river frontage, and multiple put-in and take-out points for tubers along the way. The park rents tubes, or you can bring your own. It also has one designated place for paddlers to park and put-in. Bring your own canoes or kayaks or rent locally. Seeing the eclipse while floating in a tube would be awesome!

You can also camp, hike, bike, geocache, spot turkeys and other wildlife, and marvel at the stars. The park sports 22.7 miles of trails ranging from easy to difficult, and crosses river bottoms, steep ridges, and wooded areas in between. The rugged back country trails offer solitude even on the busiest weekend. Whether you'd like a moderate hike or a more challenging mountain bike ride, expect great views and a very different experience than in the park's lowlands. The park even loans out GPS units for use in the park.

Kickapoo Cavern State Park

Quick Facts	Number of Sites	People allowed per site	Restrooms w/showers available	Distance in	Nightly Fee
Campsites w/ electricity (no RVs over 36')	5	6	Yes	Drive-in	$20
Campsites w/ water (tent or small campers only)	10	6	Yes	Drive-in	$12
Group campsite	1	30	Yes	Drive-in	

Kickapoo Cavern is located about three hours west of San Antonio and about four hours west-southwest of Austin.

Local Eclipse Circumstances			
Kickapoo Cavern State Park			
Annular Eclipse Saturday October 14, 2023		Total Solar Eclipse Monday April 8, 2024	
Duration of Annular Eclipse	3m 13.2s	Duration of Total Eclipse	3m 48.6s
Eclipse Start	9:21 AM	Eclipse Start	11:12 AM
Annularity Start	10:49 AM	Totality Start	12:29 PM
Annularity End	10:52 AM	Totality End	12:33 PM
Eclipse End	12:28 PM	Eclipse End	1:52 PM

Bring a sense of wonder and your spirit of adventure to Kickapoo Cavern State Park. Head west about 150 miles from San Antonio to explore this lightly-developed park with its many caves, birds, bats, trails and more. There is no trash service, so you need to pack out whatever you bring in.

The park has 20 known caves; the two largest are Kickapoo Cavern about 1,400 feet long and Stuart Bat Cave. Unauthorized entry into caves is not allowed to protect the caves and the creatures that live in them;

however, guided tours are available every Saturday at 1 p.m.

Three very different natural zones intermingle here, creating a patchwork of plant and animal life. Sprawling live oaks of the Edwards Plateau, cacti of the Chihuahuan Desert, and thorny shrubs of the South Texas plains coexist in the park. This mixed plant life provides habitat for many animals. Over 240 migrant and resident bird species have been sighted in the park. Borrow a pair of binoculars at headquarters and visit the bird blind to see who is stopping by for a drink.

Since both eclipses are during the bat's regular season, March–the end of October, you can expect to see thousands of Mexican free-tailed bats leave the cave each evening at dusk to hunt for their insect prey.

Garner State Park

This park can be found about two hours west of San Antonio and four hours west-southwest of Austin.

Garner State Park is located on the southwestern edge of the Edwards Plateau. The park is part of a unique sub-region known as the Balcones Canyonlands. Edwards limestone uplifted

Quick Facts	Number of Sites	People allowed per site	Restrooms w/showers available	Distance in	Nightly Fee
Campsites w/full hookup	12	8	Yes	Drive-in	$35
Campsites with Electricity	176	8	Yes	Drive-in	$26
* Two locations New and Old Garner, River Frio (20/30/50 amp available at some sites)					
Campsites w/ Electricity: New Garner, (30 amp only)	35	8	Yes	Drive-in	$22
Campsites with Water Tent or small campers only	75	8	Yes	Drive-in	$15
Screen Shelters Old Garner (no RVs or Tents)	21	8	Yes	Drive-in	$35
Screen Shelters New Garner (no RVs or tents)	16	8	Yes	Drive-in	$30
Cabins with Fireplace (no RVs or tents)	13	4-6	Bathroom and Kitchen in Cabin	Drive-up	$150 ($10 fee per person >4) Cleaning fee required
Group Site w/ dining hall and 5 bunkhouses - no RVs or Tents	1	40	Yes	Drive-up	$400 deposit and cleaning fee required
Group Hall w/ sinks and commercial stove	1	64	Yes	Drive-up	$75

millions of years ago, creating steep canyon walls and some of the most spectacular views in the Texas Hill Country. Deep canyons, crystal clear streams, high mesas, and carved limestone cliffs characterize this dramatic terrain.

Local Eclipse Circumstances Garner State Park			
Annular Eclipse Saturday October 14, 2023		Total Solar Eclipse Monday April 8, 2024	
Duration of Annular Eclipse	4m 47.7s	Duration of Total Eclipse	4m 26.2s
Eclipse Start	9:22 AM	Eclipse Start	11:13 AM
Annularity Start	10:49 AM	Totality Start	12:30 PM
Annularity End	10:54 AM	Totality End	12:34 PM
Eclipse End	12:30 PM	Eclipse End	1:53 PM

Formed in the Cretaceous age, which lasted from 138 million to 63 million years ago, the Glen Rose formation is a collection of limestone, shale, marl, and siltstone beds, formed from sediment along the shifting margins of an ancient sea. Dinosaurs roamed there and left footprints in the sands. The sea then spread over the rest of Texas, depositing the Edwards formation (limestone) over the Glen Rose beds. This layering—Glen Rose below, Edwards above— is found throughout this area.

The biggest draw at the park is the Frio River and its 16 miles of scenic hiking trails and camping options. With 2.9 miles of Frio River winding through 1,774 acres of scenic Hill Country terrain, you can swim in the Frio River, float its waters on an inner tube, or operate a paddle boat. You can also camp, study nature, picnic, canoe, fish, play miniature golf, geocache, and ride bikes. Visitors can rent paddle boats, kayaks, and inner tubes, as well as tables, barbecue pits, heaters, and fans.

Overnight visitors can stay in screened shelters, cabins, or campsites. Large groups can rent the screened shelter or group campsite. The park's concessionaire sells meals and snacks during the busy season and rents the pavilion in the off season.

Lost Maples State Natural Area

Quick Facts	Number of Sites	People allowed per site	Restrooms w/showers available	Distance in	Nightly Fee
Campsites w/ electricity	28	8	Yes	Drive-in	$20
Primitive campsites (hike-in)	50	6	Hike to a composting toilet	Varies	

Lost Maples State Natural Area and Park are located about three hours southwest of Austin and two hours northwest of San Antonio.

Local Eclipse Circumstances Lost Maples State Park			
Annular Eclipse Saturday October 14, 2023		Total Solar Eclipse Monday April 8, 2024	
Duration of Annular Eclipse	4m 59.4s	Duration of Total Eclipse	4m 25.9s
Eclipse Start	9:22 AM	Eclipse Start	11:13 AM
Annularity Start	10:49 AM	Totality Start	12:30 PM
Annularity End	10:54 AM	Totality End	12:35 PM
Eclipse End	12:30 PM	Eclipse End	1:54 PM

Once upon a time, North America was covered in maple trees, but as the last glacial period ended, maple trees migrated north away from the hotter climates except for small pockets like the ones found in Lost Maples State Natural Area. The fall foliage of this large, isolated stand of uncommon Uvalde bigtooth maples is spectacular. The foliage changes color the last two weeks of October through the first two weeks of November, depending on the weather.

In 2020, the fall colors started around October 15. This would be a great place to see the accompanying Annular Eclipse on October 14, 2023. Also note the Annular and Total Eclipse Times. They are within 30 seconds of each other, and the park is at centerline for both eclipses.

Again, you'll want to be sure to book as early as possible as the park tends to fill up for the fall foliage change and only has space for 250 cars. It will only be more in demand for eclipse viewing.

You'll want to come back for the April 8, 2024 Total Solar eclipse, too. While the maples will not be in their full fall beauty, the park is spectacular year-round. During your visit, you'll see abundant wildflowers, steep canyon walls, and the scenic Sabinal River. The park has over 10 miles of trails to explore, including a loop that takes you along the top of a 2,200-foot cliff.

Relax and wet a hook in the Sabinal River or Can Creek. You do not need a fishing license to fish from shore or pier in a state park or natural area. This area is home to a wide variety of birds, including the endangered golden-cheeked warbler and the recently delisted black-capped vireo.

Hill Country State Natural Area

This protected area is located about two hours southwest of Austin and a little over one hour west of San Antonio.

Quick Facts	Number of Sites	People allowed per site	Restrooms w/showers available	Distance in	Nightly Fee
* Except group lodge, all Primitive campsites are walk-in (WI) or hike-in (HI), no water					
(WI) West Verde Creek	3	8	Chemical toilet	50 yards to walk-in	$12
(WI) Comanche Bluff	3	8	Vault toilet nearby	25 yards to walk-in	$12
(WI) Chaquita Falls	4	8	Chemical toilet near	75 yards to walk-in	$12
(HI) Butterfly Springs	2	4	None	~ 1 mile hike from parking	$10; First come only, no reservations
(HI) Hermit's Shack	3	4	None	1.25 mile from parking	$10 – First come only, no reservations
(HI) Wilderness	5	4	None	2.5-mile hike from the closest parking	$10 – First come only, no reservations
Group Lodge w/ 5 electric only RV hookups	12 in lodge	Up to 30 total w/ others in RVs and tents nearby	1.5 Bathrooms with Toilet and shower	Drive-up	$300 2 Queen Bed 5 Twin Beds

For those who like to get away from it all or prefer to experience eclipses with just close friends and family, the Hill Country State Natural Area may be right for you. So close to two of Texas' larger cities, the park hosts over 5,000 acres of rugged canyons, scenic plateaus, and tranquil creek bottoms on a former ranch northwest of San Antonio. With only primitive campsites in the park, some with a substantial hike in, you can escape from the bustle of modern life to a more relaxed time and place.

Local Eclipse Circumstances Hill Country State Natural Area			
Annular Eclipse Saturday October 14, 2023		**Total Solar Eclipse Monday April 8, 2024**	
Duration of Annular Eclipse	4m 56.5s	Duration of Total Eclipse	4m 04.5s
Eclipse Start	9:22 AM	Eclipse Start	11:13 AM
Annularity Start	10:50 AM	Totality Start	12:31 PM
Annularity End	10:55 AM	Totality End	12:35 PM
Eclipse End	12:31 PM	Eclipse End	1:54 PM

If you'd like to bring a bigger group with you, there is a lodge that sleeps 12 with room for five RVs and tents on the grounds – up to 30 total.

The Hill Country State Natural Area is a scenic mosaic of rocky hills, seasonal flowing springs, oak mottes, grasslands, and canyons. The terrain ranges from flat, broad creek bottoms to steep, rocky canyons up to 2,000 feet high. West Verde Creek provides a small oasis against the backdrop of rocky hills and limestone cliffs. The natural area provides habitat for a great variety of birds, mammals, and amphibians, as well as various reptiles. Several different plant communities grow here, too.

Hill Country State Natural Area offers primitive camping, backpacking, nature watching, and multi-use trails for hikers, mountain bikers, and horseback riders. The trails range from easy to challenging, from one-mile strolls to miles-long rambles. The easy Heritage Loop takes you past remnants of the former ranch. The West Peak Overlook is a staff favorite, with a steep climb leading to expansive views of the western Hill Country.

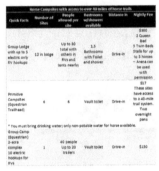

You can also bring your own horses to explore the natural area (you must present proof of current Coggins). The equestrian campsites have access to 40 miles of horse trails. The day-use equestrian area is next to headquarters and has a water trough, hitching posts, and a flush toilet. You can also stay overnight at an equestrian campsite. To conserve water, they do not have horse-washing stations.

Old Tunnel State Park

One hour northwest of San Antonio and one and a half hours west of Austin, Old Tunnel State Park is the smallest state park in Texas, consisting of only 16.1 acres of land. Despite its small size, there are a variety of recreational and wildlife-viewing opportunities. The abandoned railroad tunnel is home to up to 3 million Mexican free-tailed bats (Tadarida brasilienses mexicana) and 3,000 cave myotis (Myotis velifer) from May through October.

Local Eclipse Circumstances Old Tunnel State Park				
Annular Eclipse Saturday October 14, 2023			**Total Solar Eclipse Monday April 8, 2024**	
Duration of Annular Eclipse	3m 30.7s		Duration of Total Eclipse	4m 15.1s
Eclipse Start	9:22 AM		Eclipse Start	11:15 AM
Annularity Start	10:51 AM		Totality Start	12:32 PM
Annularity End	10:54 AM		Totality End	12:37 PM
Eclipse End	12:31 PM		Eclipse End	1:56 PM

In this park you can enjoy hiking, bird-watching, and general wildlife viewing on the half-mile nature trail. The trail opens daily at sunrise and closes at 5 p.m. You'll need to bring drinking water, as no water is available in the park. Picnic tables are available. Restrooms are only open in the evenings during bat season (May to October). To minimize disturbance to the bat colony and for your safety, you must stay on the designated trail, and you may not approach the tunnel. Due to the sensitive nature of this park, the State of Texas does not allow camping, pets, or smoking.

This park is also located in Fredericksburg, one of the state's most famous German communities. The city's German heritage is on display at the Pioneer Museum, which features settlers' homesteads and artifacts. In the nearby town square, Marktplatz, the Vereins Kirche is a replica of a 19th-century German church that once stood in the city. The vast National Museum of the Pacific War features WWII exhibits, including a recreated combat zone. Lots to do in addition to viewing an amazing eclipse.

Future Texas Eclipses

As I mentioned earlier, eclipses in Texas are rare. After 2024, the next few Total Solar Eclipses coming over Texas will just "nick" the north and south ends.

The USA Total Solar Eclipse of August 12, 2045 will clip across the top right corner of the panhandle. Only the town of Perrytown at the crossroads of Hwy 15 and Hwy 83 will get about 1 minute, 35 seconds of Total Solar Eclipse. Our neighbor state, Oklahoma, will get most of the glory for this one.

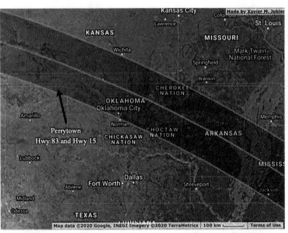

August 12, 2045 Total Eclipse Path Nicking the Top Right Corner of the Texas Panhandle

The March 30, 2052 eclipse will cross Mexico, but its only location in Texas will be Brownsville on Texas' most southern coast.

The May 11, 2078 Total Solar Eclipse will cross a few more Texas cities, but not many. Only McAllen, Harlingen, Brownsville, and South Padre Island will be in the path.

As you can see, if you live in Texas and you want an easy way to see your first eclipse, make plans to be in the path on October 14, 2023 or April 8, 2024 or you could end up waiting more than 50 years to see it again. And even then, you'll have to do some traveling just to get close for a very short period of time.

CHAPTER SIX

SOLAR ECLIPSE 101

All of us on Earth are lucky to be able to witness/experience Annular and Total Solar eclipses. Because of an interesting quirk of our solar system and orbital dynamics, we are witness to an amazing astronomical event. During the right times in the Moon's orbit around the Earth, the Moon casts a shadow on Earth, blocking a portion or all of the Sun from view despite the fact that the Sun is 400 times the size of the Moon. This is partially possible because it is also 400 times farther away,

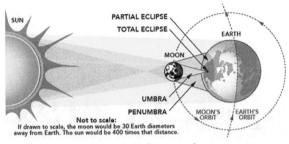

*Diagram showing the Earth-Sun-Moon
geometry of a total solar eclipse. Not to scale.
The Moon is 30 diameters away from Earth.
The Sun would be 400 times that distance. The
Sun is 400 times larger than the Moon. Drawing
credit: NASA*

We've done all this talking about solar eclipses and haven't yet taken a break to discuss what they are. Scientifically speaking, a solar eclipse occurs when the Moon orbits between the Sun and the Earth casting a shadow on the Earth. Depending on orbital dynamics, this can then block out a portion (Partial or Annular Eclipses) or all of the Sun (Total Eclipse) from our view from a specific, narrow band of activity on the Earth.

Eclipses only occur in a New Moon phase. You might think we'd have a Total Solar Eclipse every month during the New Moon. That does not happen for several reasons. First, the apparent size of the Sun varies during the year because Earth's orbit is not a perfect circle. It is an ellipse, like an egg. Our planet is closest to the Sun (perihelion) in early January and farthest (aphelion) in early July. So, the Sun appears about 3% larger in January than in July (not that you'd notice), which means at times it's harder for the Moon to completely cover the Sun and create a Total Eclipse.

Second, the Moon's apparent size changes due to its elliptical (like an egg) orbit around Earth. When the Moon is closest to Earth

(perigee), its apparent diameter is 14% larger than when it's farthest (apogee). When at apogee, the Moon is too small to cover all the Sun's brilliant face. At mid-eclipse, an annulus (ring) of sunlight surrounds the lunar silhouette, resulting in an annular eclipse. This will be the October 14, 2023 eclipse.

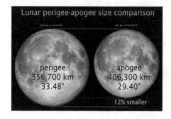

The moon will again pass directly between Earth and the Sun — but it will not quite completely cover the solar disk, instead turning it into a thin ring of fire. This annular (Latin for ring-shaped) eclipse will be visible within a roughly 125-mile-wide path from Oregon to Texas and on into Mexico, Central America, and South America.

When near perigee (closest to the Earth), the Moon can easily cover the entire solar disk and create a Total Solar Eclipse. This will be the eclipse of April 8, 2024, a 124-mile-wide path crossing up from Mexico, across Texas, through the Midwest states, into the northeast, and up into Canada.

Third, the Moon's orbit is tilted about 5° to the plane of Earth's orbit around the Sun, referred to as the Sun-Earth Ecliptic plane. As the Moon orbits the Earth, it is sometimes above or below the Sun-Earth Ecliptic plane, and the shadow of the Moon is cast into space. Only when the Moon's orbit crosses the ecliptic plane, the crossover point is called a node, does the Moon's shadow fall on Earth, creating an eclipse. This happens about every 173.3 days, roughly every six months. Depending on other factors, sometimes it's a partial eclipse, sometimes an annular, and sometimes a Total Solar Eclipse.

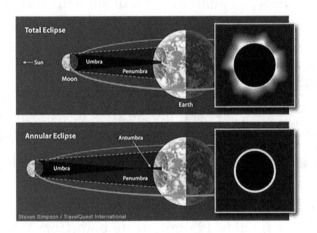

The Moon orbits around the Earth every 29.53 days, which is known as a synodic month. The time it takes for the Moon to cross the Sun-Earth orbital plane at its orbital nodes twice is called a draconic month 27.21 days. In addition, for a Total Solar Eclipse, the Moon must be at perigee, that is closer to the Earth in its elliptical orbit. The lineup of all these required factors and alignments— a new Moon, crossing the Sun-Earth orbital plane, and at perigee— is rare.

But it isn't as rare as most people think. While eclipses might be rare in a single spot, Total Solar Eclipses usually occur every 11-18 months somewhere on the planet. There are exceptions as between the August 17, 2017 USA Eclipse and the July 2, 2019 Chile/Argentina Eclipse, a gap of about 23 months. There was also a gap of about 28 months between the November 13, 2012 Australia Eclipse and the March 20, 2015 North Atlantic Eclipse.

The complex math does not end there but be careful. This can get pretty complicated. The nodes (where the Moon crosses the Sun-Earth ecliptic plane) slowly shift (precess) westward, which means the months in which eclipses take place slowly change as the years pass. This also affects the type of eclipse that occurs: currently long annulars are more likely in January, long totals in July.

Finally, after 6,585.32 days (18 years, 11 days, 8 hours), the entire eclipse cycle repeats. This is known as the Saros cycle. When two eclipses are separated by a period of one Saros, the Sun, Earth, and Moon return to approximately the same relative geometry, and a nearly identical eclipse cycle will occur (though the eclipse path will be shifted west by eight hours — one third of Earth's rotation). This cycle was known to the Chaldeans as early as 600 BC. The Saros arises from a natural harmony between three of the Moon's orbital periods (circa 2000 CE):

- Synodic Month (New Moon to New Moon) = 29.530589 days = 29d 12h 44m 03s
- Anomalistic Month (perigee to perigee) = 27.554550 days = 27d 13h 18m 33s
- Draconic Month (node to node) = 27.212221 days = 27d 05h 05m 36s

One Saros is equal to 223 synodic months. However, 239 anomalistic months and 242 draconic months are also equal to this same period (to within a couple hours)!

- 223 Synodic Months = 6585.3223 days = 6585d 07h 43m
- 239 Anomalistic Months = 6585.5375 days = 6585d 12h 54m
- 242 Draconic Months = 6585.3575 days = 6585d 08h 35m

Any two eclipses separated by one Saros cycle share very similar paths. They occur at the same node with the Moon at nearly the same distance from Earth and at the same time of year. Because the Saros period is not equal to a whole number of days, its biggest drawback is that subsequent eclipses are visible from different parts of the globe. The extra third of a day displacement means that Earth must rotate an additional ~8 hours or ~120º with each cycle. For solar eclipses, this results in the shifting of each successive eclipse path by ~120º westward. Thus, a Saros series returns to about the same geographic region every 3 Saroses (54 years and 34 days).

Yes, it can get confusing, so let's pull back for an example. The 1999 and the 2017 Total Solar Eclipses are of the same Saros 145. Their paths are similar with a long path crossing over Earth. For the August 11, 1999 eclipse, the path went from the North Atlantic off the coast of North America, across Europe, the Mid-East and on into India, Max Duration was 2m23s in Romania at 11:03UT. The August 17, 2017 Total Solar Eclipse started in the middle of the Pacific Ocean, crossed the USA from Oregon to South Carolina, ending in the Atlantic Ocean, Max Duration was 2m40s in Kentucky at 18:25UT. The next eclipse of Saros 145 will be the September 2, 2035 eclipse crossing a large path from the Xinjaing Province in China, across North Korea, Japan, and into the Pacific Ocean, Max Duration in the middle of the Pacific will be 2m54s at 01:55UT.

You'll note that the Totals of Saros 145 are getting longer, that is because the first Total Solar eclipse of this Saros started in 1927 at 50s Max Duration, this will continue until the longest eclipse of this Saros on June 25, 2522 with a max of 7m12s. After that, the duration of the total eclipses of Saros 145 will drop until the last Total Solar eclipse of this cycle ends in

Antarctica on September 9, 2648 with a duration of 2m43 seconds.

The October 14, 2023 Annular eclipse belongs to the Saros Cycle of 134. There were only eight Total and 16 Hybrid Solar eclipses in this cycle, ending with 7s of Totality in Africa on June 27, 1843. Annular eclipses in this cycle will continue until May 21, 2384.

The April 8, 2024 Total Solar eclipse belongs to the Saros Cycle of 139. The next Total Solar eclipse of this cycle will cross Indonesia and the Philippines on April 20, 2042. In Saros 139 there are 43 Total, 12 Hybrid Solar eclipses, with the cycle ending with 35s of Totality on Mar 26, 2601.

We are so fortunate to see these two Texas solar eclipses of two different Saros Cycles within six months of each other.

With my history of chasing Total Solar Eclipse, I'm now seeing the second eclipses of Saros Cycles, and expect to live long enough to see third eclipses in some of these Saros Cycles.

Saros Cycle	Saros' Eclipses Experienced		Saros Eclipses Planned	
	1st Eclipse	2nd Eclipse	2nd Eclipse	3rd Eclipse
120	Mar 20, 2015 (North Atlantic)		Mar 30, 2033 (Last Total of the Saros)	
126	Aug 1, 2008 (China)		Aug 12, 2026 (Spain/Iceland)	Aug 23, 2044 (Last Total of the Saros)
127	June 21, 2001 (Africa)	July 2, 2019 (Chile)		July 13, 2037
129	Apr 8, 2005 (Pacific)		Apr 20, 2023 (Australia/ Indonesia)	Apr 30, 2041
130	Feb 26, 1998 (Caribbean)	Mar 9, 2016 (Indonesia)		Mar 20, 2034
133	Nov 13, 2012 (Australia)		Nov 25, 2030	Dec 5, 2048
136	July 11, 1991 (Mexico)	July 22, 2009 (China/Japan)		Aug 2, 2027 (Europe/ Mid-East)
139	Mar 29, 2006 (North Africa)		Apr 8, 2024 (Texas)	Apr 20, 2042
142	Dec 4, 2002 (Australia)	Dec 14, 2020 (South America)		Dec 26, 2038
145	Aug 11, 1999 (Europe)	Aug 21, 2017 (USA)		Sept 2, 2035
146	July 11, 2010 (S. America)		July 22, 2028	Aug 2, 2046
152	Nov 3, 2003 (Antarctica)		Dec 4, 2021 (Antarctica)	Dec 15, 2039

An interesting thing I noted when doing the above table of the Saros Total Solar Eclipses I've experienced. In my lifetime, I'll see the last Total Solar Eclipses of two Saros cycles, Saros 120 and 126. Also, if I keep up my plan to live to 104, I'll be able to see the first Total Solar Eclipse of Saros 149 on April 9, 2043 in Northern Russia starting a new run of Total/Hybrid Solar Eclipses that will run until November 3, 2385 with the last Hybrid eclipse.

eclipse cycle of 10,571.95 days (about 29 years minus 20 days). Two eclipses in an inex series take place alternately at the one and the other node (unlike eclipses in a Saros series). Therefore, an eclipse in the northern hemisphere of the Earth will be followed, after one inex, by an eclipse in the southern hemisphere. I've only seen two eclipses in one inex

Series, the July 11, 1991 eclipse in Mexico and the December 14, 2020 eclipse in South America. For more on inex, I recommend Fred Espenak's eclipsewise.com pages on the subject.

For a list of all the eclipses in your lifetime, I recommend following the big three names of eclipse sites and maps – Fred Espenak, Xavier Jubier and Michael Zeiler.

Together with Jean Meeus, Fred Espenak calculated and cataloged the eclipses occurring on the planet during the 5,000-year period from 1999 to about 3,000 CE. These are listed in the NASA Technical Publication TP-2006-214141 and available in a book titled *Five Millennium Catalog of Solar Eclipses: -1999 to +3000*, available on Amazon.

From their calculations, we know that in the 5000 years from -1999 BCE to 3000 CE, the Earth will experience 11,898 eclipses of the Sun, 4,200 partial eclipses, 3,956 annular eclipses, 3,173 total eclipses and 569 hybrid eclipses. When you consider that only 3,956 Annular and 3,173 Total Eclipses in 5,000 years occur, and here in Texas we get one of each within six months, it's a very special event.

I consider Fred Espenak the "father" of the eclipse websites. During his tenure with NASA, he created and maintained NASA's eclipse webpages which can be found here: https://eclipse.gsfc.nasa.gov/solar.html. He did eclipse details before the internet. I had his 1991 NASA book of the eclipse that led me to the location where I saw my first TSE just north of Puerta Vallarta,

Mexico. I own his compendium Five Millennium Catalog of Solar Eclipses: -1999 to +3000.

He now hosts the sites mreclipse.com and eclipsewise.com, too. His sites are full of descriptions, photos, orbital dynamics, and calculations of Lunar and Solar Eclipses. He currently lives in a dark sky astronomical development in Arizona.

Xavier M Jubier is the inventor/developer of solar/lunar eclipse Google Maps and Google Earth files using Fred Espensak and Jean Meenus's data. He is also the creator of Solar Eclipse Maestro, Lunar Eclipse Maestro, and Mercury Venus Transit Maestro. If you are a serious eclipse photographer, you will get to know these programs. He hosts an Interactive Google Maps site called Lunar and Solar Eclipses:

http://xjubier.free.fr/en/site_pages/Solar

EclipsesGoogleMaps.html
https://youtu.be/e1nYi5gO70M

Fortunately for me, there was a discount version at the $2,000 per seat range that my husband and I joined out of Denmark. Our video, can be seen here: https://fb.watch/8zpIgomf6i/

Michael Zeiler is a cartographer who has taken the work of Xavier Jubier and Fred Espenak and created some beautiful and amazing maps. He also provides lots of eclipse swag to decorate your home, yourself, and carries many eclipse books. Go to

Eclipse-maps.com and look around. I have Mike's map of Total Solar Eclipses from 2010-2060 on my wall and incorporated into my vision board. More on that in a bit, but first, I want to prepare you for what to expect when you attend one of the upcoming Texas eclipses, maybe for the first time.

Chapter Seven

WHAT TO EXPECT

Annular Eclipse

The first event coming up in Texas is the October 14, 2023 Annular Eclipse. This is a great practice run for the April 8, 2024 eclipse. Annular eclipses are not as dramatic as Totals, but interesting things still happen. For this eclipse, the Moon is in apogee (remember, that's furthest from the planet and therefore smaller in the sky and less able to block the Sun).

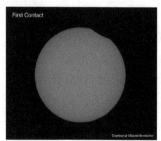

It's important to protect your eyes at all times during the annular eclipse.

The Partial Phase will start with a small bite out of the Sun. This is known as "First contact."

As the Moon continues to progress across the face of the Sun, more of the Sun's face will be obscured by the Moon. You'll also notice that the light does not appear to change or decrease as you might expect. Noticeable light reduction does not occur until the Moon covers about 50% of the Sun.

This is a good time to find a tree or pull out a colander and see the odd half-moons in the holes as the sunlight shines through. These shadows are different from anything you might expect.

At about 80 percent, you'll start noticing that the light looks funny. Shadows appear 'sharper.' If you are on a beach where there is sand, the sand looks more 'sparkly.' Through your eclipse shades or solar filters, the Sun will appear to be a small crescent.

When the eclipse gets to about 95% coverage, a slender sliver of a crescent, the light reduction is finally noticeable. It's at this point I usually start getting an eerie feeling. What was a bright, sunny day now is almost like a sepia-toned photograph with muted colors. You'll also notice the temperature dropping. It'll likely drop several degrees or more. If you are seeing this from a normally non-windy location, a sudden wind may spring up.

If you are at a location with wildlife, you'll notice the birds starting to take actions like flying to their roosting spots. Watching from the beach near Corpus Christi or Rockport, you'll likely notice the seagulls flying toward shore to find their normal roosting places. Crickets and other nocturnal insects often will start their normal nighttime chirping.

I hope any of the state parks with bats will allow visitors during both eclipses, as I am curious to see if the bats start coming out of their cave as the light diminishes.

At this point, Second Contact, the start of Annularity, the 'ring of fire' starts to appear. If you are lucky, you may catch some Bailey's Beads at the edge of the Moon as it crosses the edge of the Sun. Bailey's Beads are formed when the last light of the Sun shows through the mountains and valleys of the Moon. They appear as a string of small pearls along the edge of the Sun-Moon crossing. You can see them as early as almost one minute before totality, but they are usually only noticeable a few seconds before totality. Bailey's Beads were named after English astronomer Sir Francis Bailey in 1836. Remember – at this point you should still be looking through filters to protect your eyes.

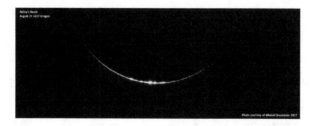

The Moon will continue to cross the face of the Sun, and while doing so, create the 'ring of fire' effect. This effect will last for up to 5 minutes 02 seconds on beaches near Corpus Christi, the length of time of annularity will depend on your location in the path.

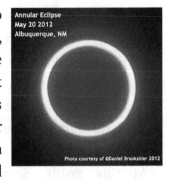

At Third contact, the annularity will end as the Moon continues its travels across the face of the Sun. Look again for Bailey's Beads just before third contact as the edge of the Moon again crosses the edge of the Sun.

At Fourth contact, the eclipse is complete, the Moon moves completely off the face of the Sun.

Remember at all times during an annular eclipse to wear proper eye protection.

Total Solar Eclipse

From first contact to the 99% obscuration, the effect of a Total Solar Eclipse such as what Texas will experience on April 8, 2024 will be similar to an annular eclipse. During the Total Solar eclipse, just before second contact, things become more dramatic and amazing.

The Bailey's Beads, because of the darkness of the event, are more pronounced just before third contact.

One thing to do at this point is to look up to see the shadow approaching from the southwest. The shadow approaches swiftly, and it's easily missed if you are not looking for it. I caught a great picture of the shadow in the sky during the August 21, 2017 eclipse in Oregon.

Shadow Bands

As the last rays of the sun begin to disappear, an often-seen effect is referred to as shadow bands. Many umbraphiles will bring a white bedsheet or white plastic tarp to a site and lay it on the ground hoping to catch the shadow band effect. When it's effective, a series of "shadows" appear to ripple across the ground or other objects, as if you are at the bottom of a pool and you see the sunlight ripple.

Another neat effect from the advancement of technology is that these shadow bands are now being captured with the new 4K cameras. In this picture from the December 14, 2020 Solar Eclipse in Fortin, Patagonia, Argentina, by Chris Chotas Alexander, the slightly vertical lines you can see across the eclipse are shadow bands visible on the cloud which blew in just as totality began. While getting a cloud at totality is usually a disaster, in this case, the thin cloud exposed the shadow bands as it moved in front of the eclipse and in addition, it created a circular rainbow around the eclipse, creating this dramatic shot.

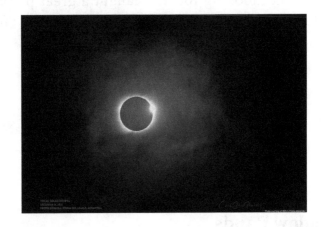

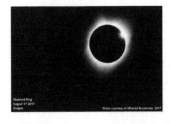

When the last ray of sunlight is still seen through the mountains/valleys of the Moon's edge, it creates a diamond ring effect as seen here. I like to say that my husband gets me a 'diamond' ring as big as the moon when we travel for eclipses together.

Second Contact starts when that last ray of sunlight disappears behind the Moon. If you are with a group of experienced umbraphiles, you'll often here a shout 'Filters off." Then the whoops, oohs, and ahs begin.

The corona appears to be shimmering, like angel's wings extending from around the Moon, reaching out into space. As you focus on the view and your eyes adjust, you'll begin to see red edges around the Moon. These are prominences, large, bright features extending outward from the Sun's surface. To us on Earth, they appear to be red flames shooting out from behind the Moon.

I love this description from Mabel L Todd, a writer of *Total Eclipses of the Sun* (1894), who followed her astronomer husband around the world, "... flashes the glory of the incomparable corona, a silvery, soft, unearthly light, with radiant streamers, stretching at times millions of uncomprehended miles into space, while the rosy, flame-like prominences skirt the black rim of the Moon in ethereal splendor."

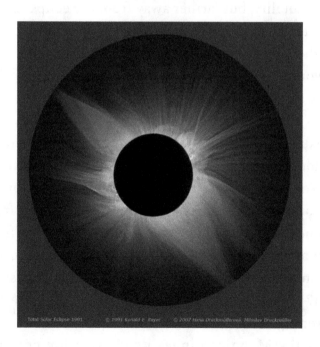

The extension of the corona fluctuates with the Sun's approximate eleven-year solar cycle. Some seem to extend out to space, some are tight around the Moon. Luckily for us, the Sun will be close to Solar

Maximum in 2024, so you can expect to see an extensive corona streaming from behind the Moon as shown in the above composite image, photo taken by Ronald F Royer, at the July 11, 1991 Total Solar Eclipse, with the composite image processed by Hana Druckmülleroviá and Miloslav Druckmüller.

In the darkness of the Total Eclipse, look for planets and stars. A bright star, Regulus, in Leo, should be close and just to the left of the eclipsed Sun, with Mercury further out to the left and Mars to the right. Venus should be visible a few minutes before and after totality, but farther away from the eclipsed Sun.

After a minute, take the time to look around you at the 360-degree sunset. This is the horizon from beyond the shadow, where the sun is still shining, but only as a thin crescent.

As the end of the total eclipse approaches, the northwest sky, which is the direction from which the shadow first approached, will get noticeably brighter.

Words, pictures, and videos can never convey a Total Eclipse experience fully, but stories from friends highlighted by some photos, videos, or films can almost instill the anticipation, enthusiasm, and excitement. Talk with those who saw the August 21, 2017 Total Solar Eclipse. It is likely they are not planning to miss this one.

As the Moon continues its progression entering third contact, you might see the diamond ring flash. Put your filters back on at this point. Bailey's Beads will appear again, until finally, a crescent of the Sun grows as you watch safely through your filters. Watch the shadow race away from you and look for shadow bands.

The Moon will continue its progression, and as the last sliver of the Moon crosses off the Sun, this is Fourth Contact. After totality, more than an hour of partial eclipse remains, as the Moon now moves slowly off the Sun. Many begin packing up before fourth contact. Others with cameras

continue trying for end-to-end shots of the partial, for that lovely composite picture treasured by many umbraphiles.

At this point, many ask the question, as I did after the 1991 Total Solar Eclipse, "When and where is the next one?"

The next one for Texas, as I mentioned above, will not be until August 12, 2045, and only for a small part of the panhandle. The next one for the USA will be August 23, 2044 and visible only from Montana and North Dakota.

I recommend traveling out of the country for the next Total Solar Eclipse on the planet on August 12, 2026, viewable from Iceland or Spain. From my point of view, highly worth the trip.

Below is a list of the last few Total Solar Eclipses and the next few Total Solar Eclipses:

- Monday, August 21, 2017 in the United States of America
- Tuesday, July 2, 2019 in Chile or Argentina
- Monday, December 14, 2020 in Argentina or Chile
- Saturday, December 4, 2021 from Union Glacier or Patriot Hills, Antarctica
- Thursday, April 20, 2023 in Australia or Indonesia
- Monday, April 8, 2024 in Mexico or the USA
- Wednesday, August 12, 2026 in Iceland or Spain
- Monday, August 2, 2027 in Egypt or Saudi Arabia
- Saturday, July 22, 2028 in Australia

Astronomers have calculated the dates and paths for Five Millennium, from -1999 BC to 3000 AD. I must say it's such a great vacation planning list and I've been doing this long enough that I'm re-visiting some continents and locales.

CHAPTER EIGHT

PRACTICAL TIPS

B y far the biggest issue for people chasing solar eclipses is protecting your eyesight. So many people seem to doubt this dire warning until it is too late. Protecting your eyes must be a top priority. The following information is provided courtesy of the American Astronomical Society at https://eclipse.aas.org/eye-safety. The AAS has much more information than I'm including here. For information on safe use of Solar Filters for photography I recommend starting on this page and going to the relevant links.

As an avid umbraphile, I realize eye-safety is important. After 20 Solar Eclipses – 18 Total and three Annular, I'm proud to say my annual eye exams report no damage from any of the Solar Eclipses I've seen. When you follow these simple rules below, your vision is safe.

Looking directly at the Sun is unsafe except during the brief total phase ("totality") of a Total Solar Eclipse, when the Moon entirely blocks the Sun's bright face. Be sure to check the location you plan to view the eclipse is located within the roughly 124-mile-wide path of the April 8, 2024 North American total solar eclipse. Xavier Jubier's Google Map (http://xjubier.free.fr/en/site_pages/solar_eclipses/

TSE_2024_GoogleMapFull.html)   supports zooming in to street level to give you the exact details of the Solar Eclipse.

During a partial or annular (ring) solar eclipse, such as the one on October 14, 2023, there is *no time* when it is safe to look directly at the Sun without using a special-purpose solar filter that complies with the transmission requirements of the ISO 12312-2 international standard.

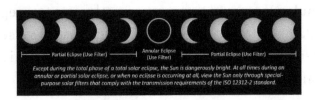

 The only safe way to look directly at the uneclipsed, partially eclipsed, or annularly eclipsed Sun is through special-purpose solar filters, such as "eclipse glasses" (example shown at left) or handheld solar viewers. Homemade filters or ordinary sunglasses, even very dark ones, are not safe for looking at the Sun. They transmit thousands of times too much sunlight. See the Reputable Vendors of Solar Filters & Viewers page for a list of manufacturers and authorized dealers of eclipse glasses and handheld solar viewers verified to be compliant with the transmission requirements of the ISO 12312-2 international safety standard for such products.

Instructions for safe use of solar filters/viewers:

- *Always* inspect your solar filter before use; if scratched, punctured, torn, or otherwise damaged, discard it. Read and follow any instructions printed on or packaged with the filter.
- *Always* supervise children using solar filters.
- If you normally wear eyeglasses, keep them on. Put your eclipse glasses on over them or hold your handheld viewer in front of them.
- Stand still and cover your eyes with your eclipse glasses or solar viewer before looking up at the bright Sun. After looking at the Sun, turn away and remove your filter — do *not* remove it while looking at the Sun.

- Do *not* look at the uneclipsed, partially eclipsed, or annularly eclipsed Sun through an unfiltered camera, telescope, binoculars, or other optical device.
- Similarly, do *not* look at the Sun through a camera, telescope, binoculars, or any other optical device while using your eclipse glasses or handheld solar viewer — the concentrated solar rays could damage the filter and enter your eye(s), causing serious injury.
- Seek expert advice from an astronomer before using a solar filter with a camera, telescope, binoculars, or any other optical device; note that solar filters must be attached to the *front* of any telescope, binoculars, camera lens, or other optics.
- If you are *inside* the path of totality on April 8, 2024, remove your solar filter *only* when the Moon completely covers the Sun's face, and it suddenly gets quite dark. Experience totality, then, as soon as the Sun begins to reappear, replace your solar viewer to look at the remaining partial phases. Note that this applies *only* to viewing without optical aid (other than ordinary eyeglasses). Different rules apply when viewing or imaging the Sun through camera lenses, binoculars, or telescopes; consult an expert astronomer before using a solar filter with any type of magnifying optics.
- *Outside* the path of totality, and throughout a partial or annular solar eclipse, you must *always* use a safe solar filter to view the Sun directly.

Note: If your eclipse glasses or viewers are compliant with transmission requirements of the ISO 12312-2 safety standard, you may look at the uneclipsed or partially eclipsed Sun through them for as long as you wish. Furthermore, if the filters aren't scratched, punctured, or torn, you may reuse them indefinitely. Some glasses/viewers are printed with warnings stating that you shouldn't look through them for more than three minutes at a time and that you should discard them if they are more than three years old. Such warnings are outdated and do not apply to eclipse viewers

manufactured since 2015, compliant with the ISO 12312-2 standard adopted that year, and in excellent condition. To make sure you get (or got) your eclipse glasses/viewers from a supplier of ISO-compliant products, see our Reputable Vendors of Solar Filters & Viewers web page.

And of course, we are talking about Texas and oil-patch country. It's common that your cousin, Dad, Mom, Grandparents, someone you know either welds for a living or a hobby. This is especially true for those in the path of the October 14, 2023 Annular eclipse crossing over Midland/Odessa. The question I get asked often is can you use welding filters? The only ones that are safe for direct viewing of the Sun with your eyes are those of Shade 12 or higher. These are much darker than the filters used for most kinds of welding. If you have an old welder's helmet around the house and are thinking of using it to view the Sun, make sure you know the filter's shade number. If it's less than 12 (and it probably is), don't even *think* about using it to look at the Sun. Many people find the Sun too bright even in a Shade 12 filter, and some find the Sun too dim in a Shade 14 filter — but Shade 13 filters are uncommon and can be hard to find. In any case, welding filters generally give a sickly green image of the Sun, whereas special-purpose solar viewers give a white, yellow, or orange image, which is much more pleasing and natural. If you really want to get a welding filter, we recommend that you buy it from a welding supply company. The AAAS has heard reports of people ordering "Shade 14" welding goggles from random online stores and receiving much lighter filters than they were promised.

What to do if you don't have filters

An alternative method for safe viewing of the partially eclipsed Sun is *indirectly* via pinhole projection. For example, cross the outstretched, slightly open fingers of one hand over the outstretched, slightly open fingers of the other, creating a waffle pattern. *With your back to the Sun, -* look at your hands' shadow on the ground. The little spaces between your fingers will project a grid of small images on the ground, showing the Sun as a crescent during the partial phases of any solar eclipse or as a ring during the annular phase of an annular eclipse. Or just look at the shadow of a leafy tree during a partial or annular eclipse; you'll see the ground dappled with crescent or ring-shaped Suns projected by the tiny spaces between the leaves.

You can also create a pinhole projector with two pieces of cardboard. Punch a hole in one piece of cardboard. With your back to the Sun, hold the cardboard up so the Sun shines through the hole and projects onto a second piece of cardboard. This was how I saw my first partial eclipse on May 30, 1984. Instructions for this can be found at:

https://www.jpl.nasa.gov/edu/learn/project/how-to-

make-a-pinhole-camera/

One of my favorite ways to see a partial eclipse is to use a colander. Yes, I'm talking about that device you use in the kitchen to strain water from your spaghetti dinner. Here is where I purchased a colander strainer spoon from a local market in Indonesia, for the March 9, 2016 Total Solar Eclipse to project the "half-moons" on the ground. Many umbraphiles make a "pinhole" sign with the date of the eclipse and take a shot of the partial through their sign to mark the occasion.

By following the above simple rules, you can safely enjoy these solar eclipses so you can see the next one.

This safety information has been endorsed by the American Astronomical Society, the American Academy of Ophthalmology, the - National Aeronautics and Space Administration, the American Academy of Optometry, the American Optometric Association, and the National Science Foundation.

Photography

I am always too enthralled with the Total Eclipse Experience to effectively operate a camera. I am lost in the streamers of the corona shooting from the Sun, looking for solar flares, the 360-degree sunset, the awe in the faces of the people around me, and my own experience of awe. This limits my ability to focus on operating mechanical devices to capture scientific measurements. Getting pictures of the eclipse has plagued scientists and hobbyists throughout recorded eclipse history. It's one of the reasons Xavier's Solar Eclipse Maestro software is popular among umbraphiles.

For those with technical expertise, I recommend Xavier Jubier's Maestro software. Just make sure your computer clock is properly set. My husband missed getting the initial first contact diamond ring in 2019 because his computer clock was a few seconds slow. My personal best was my video of the 2010 eclipse in Patagonia Argentina. My 30 second version of the video is at: https://www.youtube.com/watch?v=iU94pz68cbg

I also recommend going to Miloslov Druckmuller's Eclipse Photography Home Page http://www.zam.fme.vutbr.cz/~druck/eclipse/Index.htm

If you follow the instructions given for the 2009 TSE, you set up your photos for later digital processing as in this photo from Andreas Möller taken at the December 14, 2020 eclipse in Argentina.

Eclipse Expedition List

- Solar Eclipse Glasses / Viewers, I usually bring several pairs to share with the locals
- Solar Filters for Camera (you can make your own with compliant Solar Filter Film). I usually bring several sheets of solar film with me for sharing.
- Eye patch
- Dark clothing for the eclipse
- Black T Shirt to block out the sun
- String

- Gaffer tape
- Scissors
- Comfortable shoes, you'll find you need to do a bit of walking and usually some hauling to get to your preferred spot.
- A chair – eclipses from first contact to fourth contact last about three hours, you'll want to sit for some of it.
- A light jacket – the temperature will drop
- A hat or sun umbrella
- Sun screen

CHAPTER NINE

MAKE ECLIPSES PART OF YOUR LIFE EXPERIENCES

I live a life many perceive as glamourous and extravagant. To some extent, that is true. Justifying the cost of my hobby and other life choices through two husbands, with a plan to outlive a third, has not always been simple.

In my daily life, I have a maid to clean my home, not once but twice a week. My justification is that I hate cleaning. With my education and skill set, the time I invest in work, in learning, in rest and recovery, does more for my income and happiness than scrubbing the toilet. Especially a toilet made dirty by a lazy husband. Or, in the current case, a husband that would barf if he scrubbed a toilet. This way, my husbands can stay lazy or at least not struggle with a frail constitution, and yet I have clean toilets.

An additional benefit is I do not invest energy in nagging my spouse to help out more. I get to save that energy for work or pleasure. I'm a project manager for a tech company where I have people that I manage daily.

The second big justification in my life is my limited savings towards retirement. Rather than worry so much about retirement, I give myself the option of spending money each year chasing Total Solar Eclipses. I justify it by saying I'm taking my retirement in installments.

Part of this decision was the situation I watched with my next-door neighbors of 15 years. I moved into my starter home in 1987 in Farmer's Branch, a bedroom community in Dallas. My next-door neighbors were a lovely couple with only one vice – smoking. Except for the expense of smoking, they lived frugally. Their house, purchased in the mid-sixties, was paid off. They did not have central heat and air as they felt it was too expensive to install and not all that necessary, having grown accustomed to the Texas weather.

Things started to get tough for them in the 1990s. David lost his long-time job as an accountant because he could not make the switch to computers and compete with the younger, computer-literate generation. With few options before him, he retired except for some light consulting work when he could find it now and then.

His wife, Betty, also picked up some book-keeping to keep some income coming in. She worked mostly for the local church to which she was very devoted, which also meant her income wasn't all that high. Their dream was to get a trailer and travel the country being 'campground hosts' at the National Parks. For those who are unaware, National Parks will accept volunteer campground hosts. You commit to a month or more staying at a particular park in exchange for a 'free' camp spot as long as you 'assist' the other campers in the park.

In 1995, they did it. They got the camper/trailer. I was so excited for them. They spent a year being campground hosts a month on/month off. Then, the medical issues from decades of smoking caught up with them. After a lifetime of living frugally, watching every dime, they only had one year to enjoy the dream they'd been building toward before it all came crashing down.

A little bit about my hobby and vision boards which I mentioned earlier. When I took this hobby on with dedication after my second eclipse in 1998, I was amazed how the world shifted around me to support it. At the time, I worked as a consultant in the telecom industry, effectively looking for a contract every six months. When I started arranging my contracts to take three or four weeks off to chase eclipses, I found appreciation for my work grew by me simply being gone. The first two weeks I'd be gone, all issues that occurred on my projects would be blamed on me. During the third and fourth weeks, I'd start getting inquiries about when would I be back. My opinion is because I took off, I was missed. My advice is whatever your passion, follow it and make time for it. Your life will shift to support it.

As I'm writing this book, through the uncertainty of the prolonged Covid Pandemic crisis (umbraphiles don't like the term corona virus), I'm happy to say I'm one of the few non-Argentinians that received a special visa to enter Argentina to see the December 14, 2020 Total Solar eclipse. So many in the umbraphile community passed or were blocked from travel by the pandemic, my husband and close umbraphile friends among them. Lucky for me, there were others with photographic skills for this eclipse.

I hope in reading this, you take the lesson I learned. If you want something for yourself, but you feel you can't do it, realize you are more limited by your own fears, doubts, and lack of courage. Surround yourself with people who share the same passion. Get to know them, listen to them on how they overcame their fears. Put yourself in their shadows if you will. Let their experiences illuminate your limits and allow you to move beyond them.

I still strive to explain how the moments I spend under the shadow of the Moon affected me so deeply. Each eclipse is an experience that is both soul-expanding and humbling. I am certain of one thing, that my goal of seeing every Total Solar Eclipse has enriched my life beyond my dreams. My fellow umbraphiles are an amazing group of people. I've met interesting and inspiring people from everywhere.

I've traveled all over the world, seen and visited awesome places that I only read about in books as a child and young adult. Now, I've been there.

By chasing shadows, I've become the person I am today, a world traveler, confident that somehow, I'll get to the next shadow with my fellow umbraphile friends. I hope your experience on April 8, 2024 will inspire you to join me in the chase for Moon shadows.

LIST OF ALL THE STATE PARKS IN THE PATH

W hile I focused on providing information in this book on just the 10 Texas State Parks in both paths, I'm including Lists of State Parks in either path of the Annular Eclipse or the Total Solar Eclipse. Please remember you can get more information on each of these Texas Parks by going to the Texas State Park website: https://tpwd.texas.gov/state-parks/.

You'll need to research details of your planned eclipse outing, make reservations on the website, and find out about any special events or restrictions in the park of your choice.

The Texas State Parks Site reservations are site specific. You can reserve from one to up to five sites for the same park and time period as much as five months in advance. The five-month reservation window opens at 8 a.m. Make your reservation early, as Annular and Total Solar eclipses will draw crowds.

Also, many of the parks have vehicle limits for each site, please check with the park directly.

Day passes can be reserved for up to eight people per vehicle and up to two vehicles per arrival date. Most reservations can be made online; however, some parks or specific sites only allow reservations by phone. Check https://tpwd.texas.gov/state-parks/ for the park of your choice. Fees are collected for the total cost of the reservation, including park entrance fees. Payment is accepted from Visa, MasterCard, or Discover card.

If you are planning to extend your stay in our great state or live here, the State of Texas offers park passes for $70, a great option. You'll get a confirmation email after you make your reservation.

The Texas State Park Service also maintains a customer service center with hours Monday through Friday, 8 a.m. to 5 p.m. The center is closed on major holidays. You can contact them via email: customer.service@tpwd.texas.gov or by phone at (512) 389-8900.

Nineteen of Texas' 86 State Parks are in the Path of the Annular Solar Eclipse, bolded parks are those in the path of both eclipses.

State Park	Eclipse Start	Annularity Start	Annularity End	Eclipse End	Duration
Monahans Sandhills State Park	9:18:52 AM	10:44:37 AM	10:48:15 AM	12:22:34 PM	3m 34.9s
Big Sprint State Park	9:18:52 AM	10:44:37 AM	10:48:15 AM	12:22:34 PM	3m 37.5s
San Angelo State Park	9:20:13 AM	10:46:53 AM	10:50:21 AM	12:25:30 PM	3m 28.6s
Devils River State Natural Area	9:20:56 AM	10:48:32 AM	10:51:04 AM	12:27:22 PM	2m 32.2s
Devils Sinkhole State Natural Area	9:21:40 AM	10:48:30 AM	10:53:25 AM	12:28:48 PM	4m 55.0s
South Llano River State Park	9:21:40 AM	10:48:38 AM	10:53:16 AM	12:28:41 PM	4m 37.6s
Kickapoo Cavern State Park	9:21:41 AM	10:49:22 AM	10:52:36 AM	12:28:55 PM	3m 13.2s
Garner State Park	9:22:23 AM	10:49:42 AM	10:54:30 AM	12:30:18 PM	4m 47.7s
Lost Maples State Park	9:22:23 AM	10:49:36 AM	10:54:36 AM	12:30:15 PM	4m 59.4s
Hill Country State Natural Area	9:22:56 AM	10:50:28 AM	10:55:24 AM	12:31:19 PM	4m 56.5s
Old Tunnel State Park	9:22:56 AM	10:53:21 AM	10:54:25 AM	12:31:07 PM	3m 03.7s
Government Canyon State Natural Area	9:23:25 AM	10:51:23 AM	10:55:59 AM	12:32:15 PM	4m 36.9s
Guadalupe State Park	9:23:28 AM	10:52:21 AM	10:55:03 AM	12:32:10 PM	2m 42.1s
Honey Creek State Park	9:23:28 AM	10:52:18 AM	10:55:07 AM	12:32:11 PM	2m 48.8s
Choke Canyon State Park	9:24:48 AM	10:53:19 AM	10:58:15 AM	12:35:03 PM	4m 55.8s
Goliad State Park	9:25:40 AM	10:55:15 AM	10:58:54 AM	12:36:33 PM	3m 39.3s
Lake Corpus Christi State Park	9:25:41 AM	10:54:38 AM	10:59:36 AM	12:36:46 PM	4m 58.2s
Goose Island State Park	9:26:35 AM	10:56:24 AM	11:00:31 AM	12:38:19 PM	2m 07.6s
Mustang Island State Park	9:26:49 AM	10:56:18 AM	11:01:20 AM	12:38:52 PM	5m 01.8s

Annular Solar Eclipse of October 14, 2023 (Times CDT)

Thirty-two of Texas' 86 State Parks are in the Path of the April 8, 2023 Total Solar Eclipse, bolded parks are those in the path of both eclipses.

Total Solar Eclipse of April 8, 2024 (Times CDT)				
State Park	Eclipse Start	Annularity Start	Annularity End	Eclipse End	Duration
Kickapoo Cavern State Park	11:12:01 AM	12:29:27 PM	12:33:16 PM	1:52:42 PM	3m 48.6s
Garner State Park	11:13:00 AM	12:30:18 PM	12:32:44 PM	1:53:53 PM	4m 26.2s
Devils Sinkhole State Natural Area	11:13:23 AM	12:30:57 PM	12:32:27 PM	1:53:53 PM	3m 30.0s
Lost Maples State Park	11:13:39 AM	12:30:56 PM	12:35:22 PM	1:54:27 PM	4m 25.9s
Hill Country State Natural Area	11:13:51 AM	12:31:28 PM	12:35:32 PM	1:54:52 PM	4m 04.5s
Government Canyon State Natural Area	11:14:18 AM	12:32:37 PM	12:35:30 PM	1:55:26 PM	2m 53.0s3
South Llano River State Park	11:14:34 AM	12:32:17 PM	12:35:27 PM	1:54:54 PM	3m 10.0s
Old Tunnel State Park	11:15:17 AM	12:32:47 PM	12:37:02 PM	1:56:06 PM	4m 15.1s
Honey Creek State Natural Area	11:15:17 AM	12:33:29 PM	12:36:35 PM	1:56:18 PM	3m 06.7s
Guadalupe State Park	11:15:19 AM	12:33:29 PM	12:36:38 PM	1:56:20 PM	3m 08.8s
Lyndon B Johnson State Park	11:15:49 AM	12:33:21 PM	12:37:32 PM	1:56:35 PM	4m 12.6s
Blanco State Park	11:15:50 AM	12:33:44 PM	12:37:22 PM	1:56:44 PM	3m 37.5s
Enchanted Rock State Natural Area	11:16:00 AM	12:33:22 PM	12:37:47 PM	1:56:36 PM	4m 25.0s
Pedernales Falls State Park	11:16:30 AM	12:34:18 PM	12:38:05 PM	1:57:18 PM	3m 47.0s
Longhorn Cavern State Park	11:17:04 AM	12:34:29 PM	12:38:51 PM	1:57:38 PM	4m 22.1s
Inks Lake State Park	11:17:09 AM	12:34:32 PM	12:38:56 PM	1:57:40 PM	4m 23.8s
Mother Neff State Park	11:19:35 AM	12:37:05 PM	12:41:20 PM	1:59:53 PM	4m 15.9s
Meridian State Park	11:20:18 AM	12:37:41 PM	12:41:46 PM	2:00:11 PM	2m 57.8s
Dinosaur Valley State Park	11:20:49 AM	12:38:37 PM	12:41:35 PM	2:00:24 PM	2m 57.8s
Lake Whitney State Park	11:20:51 AM	12:38:10 PM	12:42:30 PM	2:00:46 PM	4m 19.9s
Cleburne State Park	11:21:13 AM	12:38:45 PM	12:42:21 PM	2:00:51 PM	3m 35.6s
Fort Parker State Park	11:21:28 AM	12:39:36 PM	12:42:46 PM	2:01:42 PM	3m 10.0s
Fairfield Lake State Park	11:22:29 AM	12:41:07 PM	12:43:21 PM	2:02:38 PM	2m 13.7s
Cedar Hill State Park	11:22:42 AM	12:40:07 PM	12:43:59 PM	2:02:10 PM	3m 51.2s
Purtis Creek State Park	11:23:41 AM	12:41:18 PM	12:45:17 PM	2:03:27 PM	3m 58.2s
Lake Tawakoni State Park	11:24:33 AM	12:41:50 PM	12:46:12 PM	2:03:59 PM	4m 21.5s
Tyler State Park	11:24:57 AM	12:43:18 PM	12:45:56 PM	2:04:40 PM	2m 37.9s
Bonham State Park	11:25:38 AM	12:43:14 PM	12:46:26 PM	2:04:30 PM	3m 12.5s
Cooper Lake State Park	11:25:52 AM	12:43:05 PM	12:47:24 PM	2:05:00 PM	4m 18.6s
Lake Bob Sandlin State Park	11:26:17 AM	12:43:52 PM	12:47:47 PM	2:05:37 PM	3m 54.5s
Dangerfield State Park	11:26:48 AM	12:44:59 PM	12:47:50 PM	2:06:11 PM	2m 21.3s
Atlanta State Park	11:27:52 AM	12:46:23 PM	12:48:34 PM	2:07:06 PM	2m 11.3s

GLOSSARY

The following are terms used to describe various aspects of solar eclipses. You'll find many of them used throughout this book and will encounter others on other sites and in the media as we get closer and closer to the October 14, 2023 and April 8, 2024 solar eclipses through Texas.

Credit to the American Astronomical Society.

Annular eclipse: A solar eclipse where the apparent diameter of the Moon is too small to completely cover the Sun. At mid-eclipse, the Sun appears as a blindingly bright ring surrounding the Moon.

Annularity: The maximum phase of an annular eclipse, when the Moon's entire disk is seen silhouetted against the Sun. Annularity occurs between second and third contact. It can last from a fraction of a second to a maximum of 12 minutes 30 seconds.

Antumbra: The extension of the Moon's shadow beyond the umbra. Within the antumbra, the Sun appears larger than the Moon, which is visible in silhouette. An observer standing in the antumbra sees an annular eclipse.

Baily's Beads: Caused by shafts of sunlight shining through deep valleys on the lunar limb (edge), they look like a series of brilliant beads popping on and off. They appear just prior to second contact and just after third. They're named after the English astronomer Francis Baily, who first described them during the annular eclipse of May 15, 1836.

Chromosphere: A thin, red-colored layer of solar atmosphere located just above the photosphere. It is briefly visible immediately after second contact and just prior to third.

Corona: The Sun's upper atmosphere, visible as a pearly glow around the eclipsed Sun during totality. Its shape (sometimes elongated, sometimes round) is determined by the Sun's magnetic field and is linked to the sunspot cycle.

Diamond ring: A single Baily's Bead, shining like a brilliant diamond set into a pale ring created by the pearly white corona. It's the signal that totality is about to start (second contact) or has ended (third contact).

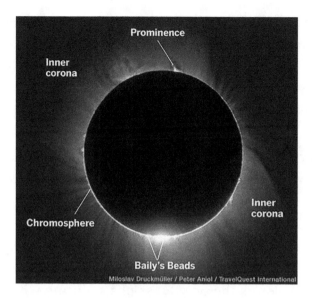

Duration: The time between second and third contact during a total or annular solar eclipse.

Eclipse magnitude: The fraction of the Sun's diameter covered by the Moon. It is a ratio of Sun/Moon diameters and should not be confused with eclipse obscuration (see next item). So, when the eclipse magnitude is 50% (50% of the Sun's diameter is covered), only about 40% of the solar surface is obscured.

Eclipse obscuration: The fraction of the Sun's surface area covered by the Moon. Do not confuse it with eclipse magnitude (see previous item). When 50% of the solar surface is obscured, the eclipse magnitude is roughly 60%.

First contact (C1): The moment when the Moon takes its first tiny nibble out of the solar disk — the beginning of the partial phase of an eclipse.

Fourth (last) contact (C4): The instant when the Moon no longer covers any part of the solar disk. This signals the conclusion of the partial phase of an eclipse.

Hybrid eclipse: A solar eclipse that begins as an annular, becomes total for a brief time, and then reverts back to an annular before it ends. This is also known as an annular-total eclipse. On rare occasions, a hybrid eclipse may begin as an annular and end as a total, or vice versa.

New Moon: The lunar phase when the Moon is located in the same direction in the sky as the Sun. New Moon is the only lunar phase during which an eclipse of the Sun can occur.

Partial eclipse: A solar eclipse where the Moon covers only a portion of the Sun. A partial eclipse precedes and follows totality or annularity, but a partial can also occur by itself. A partial solar eclipse is visible over a wider swath of Earth than is totality or annularity.

Path of Totality: The straight-line path of the Moon's shadow on the Earth as it blocks out the light of the Sun. Standing anywhere in this path is the only place the Total Solar Eclipse can be witnessed.

Penumbra: The portion of the Moon's shadow in which only part of the Sun is covered. An observer standing in the penumbra sees only a partial solar eclipse.

Photosphere: The visible surface of the Sun, which consists of a gas layer at a temperature of roughly 5,500° Celsius (10,000° Fahrenheit).

Prominence: Hot gas hanging just above the solar surface, usually appearing as a red-colored arc or filament hovering in the lower part of the corona. Prominences are quickly covered by the Moon after second contact and revealed just prior to third.

Saros: An eclipse cycle with a period of 6,585.32 days. When two eclipses are separated by a period of one Saros, the Sun, Earth, and Moon return to approximately the same relative geometry, and a nearly identical eclipse will occur (though the eclipse path will be shifted west by eight hours — one third of Earth's rotation).

Second contact (C2): The instant when the total or annular phase of an eclipse begins. For a total eclipse, this is synonymous with the disappearance of the first diamond ring.

Shadow bands: Very faint, shimmering ripples of dark and light moving across the ground, walls, or clouds. These hard-to-see bands result from atmospheric "twinkling" of the thin solar crescent just before second contact and/or just after third contact. When they occur at all (they're not reported at every total eclipse), they're best seen against a white background.

Sunspots: Dark regions on the Sun where magnetic fields are bundled together and are so strong that the flow of hot gas from the Sun's interior to the surface is inhibited. The spots appear dark because their temperature is about 1,000° Celsius (1,800° Fahrenheit) cooler than the photosphere that surrounds them.

Sunspot cycle: The number of sunspots rises and falls in roughly an 11-year cycle. During sunspot minimum, the Sun's corona appears elongated with streamers extending away from the solar equator; during sunspot maximum, the Sun's corona appears more rounded and symmetrical.

Third contact (C3): The instant when the total or annular phase of a solar eclipse ends. For a total eclipse, this is synonymous with the appearance of the second diamond ring.

Total eclipse: A solar eclipse where the apparent diameter of the Moon is large enough to completely cover the Sun's photosphere (even if only momentarily) and reveal the faint solar corona.

Totality: The maximum phase of a total solar eclipse, during which the Moon's disk completely covers the Sun. Totality occurs between second and third contact. It can last from a fraction of a second to a maximum of 7 minutes 31 seconds.

Umbra: The darkest part of the Moon's shadow. Within the umbra, the Moon appears larger than the Sun. An observer standing in the umbra sees a total solar eclipse.

Umbraphile A solar-eclipse aficionado; a person who will do almost anything, and travel almost anywhere, to see totality. Another term for an umbraphile is "eclipse chaser."

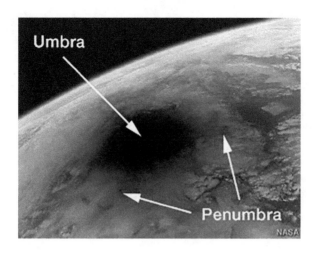

RESOURCES

Https://www.eclipse-chaser-log.com/
https://eclipse.aas.org/eye-safety
Miloslav Druckmuller -
http://www.zam.fme.vutbr.cz/~druck/
Eclipse Maps -
https://eclipse-maps.com/Eclipse-Maps/Welcome.html
Fred Espenak - http://eclipsewise.com/
Five Millennium Canon of Solar Eclipses Database
- https://mreclipse.com/
Xavier M. Jubier's interactive maps -
http://xjubier.free.fr/en
NASA's eclipse website -
https://eclipse.gsfc.nasa.gov/solar.html
Mike Nichols - https://hometownbyhandlebar.com/
Oliver Staiger - http://www.klipsi.ch/
Texas State Park website -
https://tpwd.texas.gov/state-parks/
Mike Zieler -
https://www.greatamericaneclipse.com/;
Eclipse-maps.com

Special thanks to Freepik for some of the clip art graphics used in this book - https://www.freepik.com

ABOUT THE AUTHOR

Leticia Ferrer's goal is to see every Total Solar Eclipse on the face of the planet until it's time to move on from this life. So far, she's traveled to six continents to get in the path of every Total Solar Eclipse since 1998, including a flight over Antarctica in 2003. She is one of the lucky few to catch the 2020 eclipse in Argentina. With her upcoming trip to Antarctica, she'll obtain her seventh continent under her feet. With her first eclipse in 1991 and each one since 1998, she's seen a total of 19 eclipses with only one cloud out and one lucky hole.

For the August 21, 2017 eclipse, she did a TedX presentation encouraging everyone to "Get in the Path." She enjoyed that eclipse with 22 friends and family in eastern Oregon.

She's ecstatically looking forward to the upcoming Texas Eclipses. It's her hobby coming home. In the case of the April 8, 2024 Total Solar Eclipse, she means this literally. Her parents' farm, where she grew up, is 10 miles from centerline.

CPSIA information can be obtained
at www.ICGtesting.com
Printed in the USA
LVHW080030180422
716461LV00011B/435

9 781954 373167